Hogenrich Damanik

FEM simulation of non-isothermal viscoelastic fluids

Hogenrich Damanik

FEM simulation of non-isothermal viscoelastic fluids

Numerical benchmarking

Südwestdeutscher Verlag für Hochschulschriften

Impressum/Imprint (nur für Deutschland/only for Germany)
Bibliografische Information der Deutschen Nationalbibliothek: Die Deutsche Nationalbibliothek verzeichnet diese Publikation in der Deutschen Nationalbibliografie; detaillierte bibliografische Daten sind im Internet über http://dnb.d-nb.de abrufbar.
Alle in diesem Buch genannten Marken und Produktnamen unterliegen warenzeichen-, marken- oder patentrechtlichem Schutz bzw. sind Warenzeichen oder eingetragene Warenzeichen der jeweiligen Inhaber. Die Wiedergabe von Marken, Produktnamen, Gebrauchsnamen, Handelsnamen, Warenbezeichnungen u.s.w. in diesem Werk berechtigt auch ohne besondere Kennzeichnung nicht zu der Annahme, dass solche Namen im Sinne der Warenzeichen- und Markenschutzgesetzgebung als frei zu betrachten wären und daher von jedermann benutzt werden dürften.

Verlag: Südwestdeutscher Verlag für Hochschulschriften GmbH & Co. KG
Dudweiler Landstr. 99, 66123 Saarbrücken, Deutschland
Telefon +49 681 37 20 271-1, Telefax +49 681 37 20 271-0
Email: info@svh-verlag.de

Approved by: Dortmund, TU, Diss., 2011

Herstellung in Deutschland:
Schaltungsdienst Lange o.H.G., Berlin
Books on Demand GmbH, Norderstedt
Reha GmbH, Saarbrücken
Amazon Distribution GmbH, Leipzig
ISBN: 978-3-8381-2710-1

Imprint (only for USA, GB)
Bibliographic information published by the Deutsche Nationalbibliothek: The Deutsche Nationalbibliothek lists this publication in the Deutsche Nationalbibliografie; detailed bibliographic data are available in the Internet at http://dnb.d-nb.de.
Any brand names and product names mentioned in this book are subject to trademark, brand or patent protection and are trademarks or registered trademarks of their respective holders. The use of brand names, product names, common names, trade names, product descriptions etc. even without a particular marking in this works is in no way to be construed to mean that such names may be regarded as unrestricted in respect of trademark and brand protection legislation and could thus be used by anyone.

Publisher: Südwestdeutscher Verlag für Hochschulschriften GmbH & Co. KG
Dudweiler Landstr. 99, 66123 Saarbrücken, Germany
Phone +49 681 37 20 271-1, Fax +49 681 37 20 271-0
Email: info@svh-verlag.de

Printed in the U.S.A.
Printed in the U.K. by (see last page)
ISBN: 978-3-8381-2710-1

Copyright © 2011 by the author and Südwestdeutscher Verlag für Hochschulschriften GmbH & Co. KG and licensors
All rights reserved. Saarbrücken 2011

To my father and son, Damanik,

my mother and wife, Julia and Shinta.

Acknowledgments

I would like to thank my main supervisor, Prof. Dr. Stefan Turek, for giving me once in a life time opportunity of immersing myself in the very front high performance computing knowledge as a PhD student. Thank you for your passionate, confidence, optimism and patience throughout these years in the field of numerical computation. I also thank Dr. Jaroslav Hron and Dr. Abderrahim Ouazzi for very helpful discussions and guidance on any numerical related topics which sharpen my understanding. Additionally all my colleagues and secretaries at the Institut für Angewandte Mathematik und Numerik Ls iii TU Dortmund should be thanked for their invaluable help and support during these last years, especially Sven Buijssen.

It should be acknowledged that this work has been made possible by the financial contributions from the Graduate School of Production Engineering and Logistics and by the German Research Foundation (DFG) through the collaborative research center SFB/TR TRR 30.

Finally, I would like to deeply thank my family, in particular my mother Julia and my wife Maria Shinta Hapsari, for always believing in me.

Dortmund, on Mai 2011

Hogenrich Damanik

Contents

1 Introduction — 1
- 1.1 General overview — 1
- 1.2 Viscoelastic Fluid — 4
- 1.3 Contribution of the thesis — 6

2 Governing equations — 9
- 2.1 General Newtonian flow — 9
- 2.2 Viscoelastic material model — 10
  - 2.2.1 Viscoelastic model with conformation stress tensor — 10
  - 2.2.2 Viscoelastic model with LCR — 11
- 2.3 Boundary and initial conditions — 12

3 The discrete time and space system — 15
- 3.1 The fully implicit time discretization — 15
- 3.2 The Finite Element Method — 16
- 3.3 The choice of finite element space — 18
- 3.4 Artificial diffusion stabilization — 21
- 3.5 EO-FEM stabilization — 22

4 The Numerical Solver — 25
- 4.1 Outer Newton solver — 25
- 4.2 The Jacobian matrix — 26
  - 4.2.1 Analytic Jacobian — 26
  - 4.2.2 Inexact Jacobian — 27
- 4.3 Inner multigrid solver — 29
- 4.4 Solver behavior — 30
  - 4.4.1 Nonlinear viscosity in flow around cylinder configuration — 30
  - 4.4.2 Flow around a cylinder with artificial diffusion — 32
  - 4.4.3 Viscoelastic flow in a cavity with EO-FEM stabilization — 34

5 Numerical Validation — 37
- 5.1 Flow around cylinder benchmark — 37
- 5.2 Driven cavity benchmark — 38
- 5.3 Standing vortex — 40
- 5.4 MIT Benchmark 2001 — 41
- 5.5 Planar flow around cylinder — 44

6 Applications — 51
- 6.1 Nonisothermal flow — 51
  - 6.1.1 Temperature dependent viscosity: heat exchanger — 51
  - 6.1.2 Temperature dependent viscosity: micro-reactor — 53
  - 6.1.3 Heat dissipation — 54
- 6.2 Viscoelastic-related flow — 57
  - 6.2.1 Lip vortex growth — 57
  - 6.2.2 Viscoelastic in a cavity — 59
  - 6.2.3 Non-isothermal viscoelastic flow in 4:1 contraction — 60
  - 6.2.4 Future extension — 61

7 Summary and Outlook — 65

CONTENTS

A Appendix   **67**
  A.1 Derivation of LCR in details 67
      A.1.1 Oldroyd-B in conformation tensor formulation 68
      A.1.2 Oldroyd-B in LCR formulation 69
  A.2 Derivation of conformation tensor inflow conditions 69
  A.3 Eigenvalues implementations 70
  A.4 Details of stretching in the wake region 71
  A.5 More MIT Benchmark 2001 results 72

Bibliography   **74**

1

Introduction

1.1 General overview

Fluid, in many cases, is part of our life. Our body consists of 80% of fluids, a tiny single cell of plankton consists of fluids, the earth and the atmosphere consist of a large area of fluids. Fluid is everywhere and becomes a very important element in all human aspects. Thus, it is not only interesting but also very important to explore fluid with experiments, modelling or simulations of fluid motion. This study explores viscoelastic fluid with numerical simulation based on FEM (Finite Element Method). Viscoelastic fluid is classified as non-Newtonian fluids because its stress depends nonlinearly on the deformation rate (shear rate). Unlike water which can be categorized as Newtonian fluid, viscoelastic fluid behaves very differently with respect to time scale of the flow in that the stress decreases gradually with time when maintaining a certain amount of shear rate. This leads to the well-known "stress relaxation" phenomena. Molten polymer[1] solution and white eggs are examples of viscoelastic fluids.

The essence of fluid motion is always described by a sound mathematical foundation. This is well-known as the Stokes or Navier-Stokes equation which serves as the basis of many CFD (Computational Fluid Dynamic) applications and can be given in the following form,

$$\rho \frac{\partial \mathbf{u}}{\partial t} + \rho (\mathbf{u} \cdot \nabla) \mathbf{u} = -\nabla p + \nabla \cdot \mathbf{T}, \quad \nabla \cdot \mathbf{u} = 0, \qquad 1.1$$

with $\mathbf{u}, p, \mathbf{T}, \rho$ are velocity vector, pressure, extra stress tensor and material density. More specific in this study, the Navier-Stokes equation is coupled with the energy equation given in this form

$$\frac{\partial \Theta}{\partial t} + (\mathbf{u} \cdot \nabla) \Theta = k_1 \nabla^2 \Theta + k_2 \mathbf{D} : \mathbf{D} \qquad 1.2$$

and/or with the stress equation given in this form

$$\mathbf{T} + \Lambda \frac{\delta_a \mathbf{T}}{\delta t} = 2 \eta_0 \left(\mathbf{D} + \Lambda_r \frac{\delta_a \mathbf{D}}{\delta t} \right) \qquad 1.3$$

for the simulation of non-isothermal viscoelastic fluid flow. Here, $k_1, k_2, \Theta, \eta_0, \mathbf{D}, \Lambda, \Lambda_r$ are thermal diffusivity, viscous dissipation, temperature, zero-shear viscosity, deformation tensor, relaxation time and retardation time. All required CFD (Computational fluid dynamic) techniques are implemented and realized within FEATFLOW (Finite Element Analysis and Tools for Flow problems). An introduction to the code can be found at http://www.featflow.de.

Physically, the flow of Newtonian fluids is categorized by the non dimensional Reynold number, Re, which tells us whether the flow is laminar, transient, or turbulent. In viscoelastic fluid flow, the physical phenomena is controlled by the amount of elastic property of the

---

[1] Polymer is an integrated unit of many simple molecules (monomer). More detail descriptions can be found in [46]

CHAPTER 1. INTRODUCTION

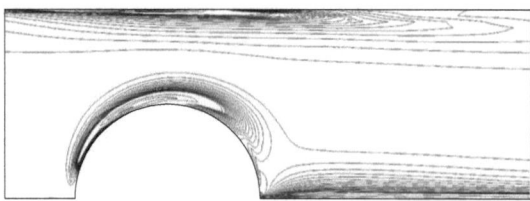

Figure 1.1: Viscoelastic flow around a cylinder with Oldroyd-B for We = 2.1.

fluid, which is a time-related parameter and describes the time-scale needed by the fluid particle to come back into its zero state (or to release its past stress memory) when the fluid motion stops. The time parameter is well-known as the relaxation time or, in non-dimensional form, the Weissenberg number, We. In the numerical context, both non dimensional numbers measure the complexity/difficulty of the corresponding numerical computations.

In the case of Newtonian fluid flows, numerical instability occurs at high Re number where the solution creates boundary layers due to problems with no-slip condition of the velocity. In contrast to this, viscoelastic fluid flow solution creates boundary layers given by property of the stress [61], not the velocity, see Fig. 1.1. The hyperbolic nature, which presents in the constitutive law, is responsible for this problem. Thus, the numerical computation to obtain a solution is already a challenging task at low Weissenberg number We < 1, precisely in the case of Oldroyd-B (later described) type of fluids where a standard Galerkin formulation is not expected to be optimal [51, 30, 45]. Yet, the remedies are available which can be reformulation of the original stress equation, numerical stabilization or/and compatible pairs of FE discretizations.

A better formulation of the constitutive laws is presented by Hulsen, Fattal and Kupfermann [41] as an alternative to the original formulation which is based on the logarithm of conformation tensor[2]. The so-called LCR (Log-conformation reformulation) is proposed to deal with the high stress gradient that occurs during numerical computation. LCR equation does not approximate directly the conformation stress tensor, but it approximates instead intermediate numerical variable from which conformation stress tensor is explicitly obtained. In this way, it is shown that the numerical stability is cured and the limit of We number increases [41], see Fig. 1.2.

This study tries to incorporate the main issue in solving viscoelastic flow problems by implementing a fully coupled monolithic approach with a consistent stabilization technique together with hanging nodes in a high order finite element frame work. The extreme development of computer resources in the last 10 years provides a wider possibility towards coupled FEM methods which was hardly done 20 years ago. This is also driven by the fact that many CFD solvers used an operator splitting approach together with a low order finite element implementation which is in fact very efficient but needs an extra care when it comes into accuracy of the solution. To the contrary, the high order finite element Q_2 towards fully coupled monolithic approach maintains highly accurate solutions. This element together with discontinuous P_1 element for the pressure space approximation satisfies the well-known LBB (named after Ladyzhenskaya, Babuška and Brezzi [34]) condition and is, without doubt from years of experiences, one of the best finite element pairs in the Stokes problem [6, 10, 70]. An example of benchmark flow around cylinder shows that this element pair can obtain an accurate direct steady solution for medium Re numbers within few Newton steps, later in chapter 5.

---

[2]The rheology of polymer depends highly on the microscopic behavior of a single molecule. Conformation stress tensor describes the macroscopic polymer behavior by neglecting interactions between single molecules [46].

1.1. GENERAL OVERVIEW

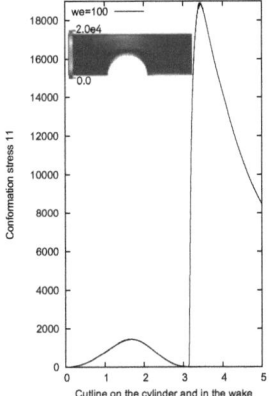

Figure 1.2: Conformation stress tensor τ_{11} with Giesekus for We $= 100$.

The aspect of numerical stabilization plays an important role when it comes to high inertia flow simulations or high elastic flows. In the first case, the convective terms inside the Navier-Stokes equation are dominant while in the second case the convective terms inside the above stress equation become also dominant. The situation is far more complicated for the second case as the total system of equations increases by the additional stress equations which introduces two new problems: i). The characteristic of the total system of equation becomes far away from the elliptic character even if the convective term is neglected from the momentum equation (creeping flow), which is a bad indication for most CFD solvers and ii). An extra compatibility constraint for the velocity-stress interpolation functions must be satisfied.

Thus, the need of a consistent stabilization technique is pronounced. There are several well-known stabilization techniques available in the CFD community, namely upwind schemes, streamline diffusion, FCT, edge-oriented stabilization (EO-FEM). This study is in favor of the last one which is based on successful experience after years working with a low order element $\tilde{Q}_1$, see [56]. In fact, the same stabilization technique can be applied to high order finite element too. One can see Fig. 1.3 which presents a lid-driven cavity flow at high Re number using higher order interpolation function Q_2, which can be hardly obtained without stabilization. For many reasons, the use of stabilization technique is still debatable. One argument can be as follows: Instead of solving the original problem, by adding stabilization term, one actually solves another problem which is similar but still not the same. This study shows later in Chapter 3 that the stabilization term can be applied only to the left hand side of the discretized system while keeping the right hand side as it is. Thus, one solves exactly the original problem with a robust convergence behavior.

The Solution method to the discrete nonlinear system arose from the discretization follows Newton iteration with line search method. The technique is well accepted as the most robust iteration technique and may give quadratic convergence. Moreover the Jacobian is computed in a 'black-box' manner which opens the door to many other constitutive material laws without having to derive each of them analytically by Frechét-derivative at the continuous level. This is done via divided difference approach. Inside one Newton step, the solution of the linearized discrete system, which is also treated in a coupled way, utilizes an efficient geometric

CHAPTER 1. INTRODUCTION

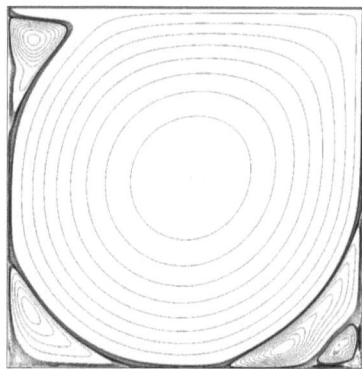

Figure 1.3: Driven cavity flow at Re = 10000 with $Q_2 P_1$ pair and robust stabilization.

multigrid solver with full prolongation and restriction operator, which is the long time work of FEATFLOW.

So, essentially new in this study is the implementation of the energy and stress equation into the Navier-Stokes equation in a fully coupled monolithic way of solving the system as well as the implementation of jump stabilization technique for the high order finite element (Edge-oriented stabilization technique) to stabilize the nonlinear term that arise from the stress and the energy equation. Since this is also an extension of the 2D Navier-Stokes solver that is used in our chair, step by step validation through benchmarking is part of the study.

This thesis is structured as follows: the first chapter introduction will be devoted to a general overview of this study, viscoelastic fluids and thesis contribution. Then, the governing equation of the corresponding problem is highlighted in the second chapter. In chapter 3, the finite element discretization as well as the time discretization techniques are presented which results in a set of discretized problems to solve. These problems are then solved iteratively by numerical solvers which is described in chapter 4. The so-called monolithic approach is presented in a way that any given viscoelastic model can be solved in the same way maintaining the same rate of convergence. Newton and multigrid solver are presented in this chapter. Then, step-by-step validation of the code with respect to different flow problems (flow around cylinder, kinetic energy in a cavity, nonisothermal flow, viscoelastic flow) is examined in chapter 5 and prototype applications of the code are presented in chapter 6. In chapter 7, the study is closed by summarizing and giving an outlook for future research.

1.2 Viscoelastic Fluid

The topic of viscoelastic fluids has been very interesting to explore since more than 100 years. It falls into the group of viscoelasticity which couples Newtonian viscous stress and elastic stress [63]. It is usually used to model molten polymer process in industry [46]. There is also group of material law that couples Newtonian viscous strain rate and plastic strain rate. This is well-known as viscoplasticity where the Newtonian viscous strain rate depends highly on the so-called 'overstress', otherwise, fluids of this group do not flow [65]. Extension of this model with extra elastic strain rate falls into the group of visco-elasto-plasticity. Other constitutive model that may be used for non-Newtonian fluid modelling is visco-hypoplasticity, mainly for modelling clay [53]. So, it is convenient to realize that this thesis focuses on viscoelasticity

1.2. VISCOELASTIC FLUID

and there are other models that can be taken into account for other research purposes.

The viscoelastic fluids generate physically remarkable effects when they are stirred, vibrated, or given a sudden external forces [11]. These effects are well-known and one of them can be seen in Fig.1.4 which is called "rod climbing". This kind of effect appears to be significant in the industrial process, where the rheology of the related material is of importance. Poisson, Maxwell and Boltzmann started the idea that any fluid has not only viscosity prop-

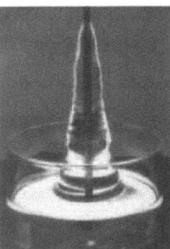

Figure 1.4: The rod climbing effect [11]. Permission from Elsevier and Prof. David V. Boger.

erty but also an instantaneous elasticity[3]. In 1929, Jeffrey wrote a constitutive law, which considers both properties. Then in 1950, Oldroyd [54] presented the constitutive law in terms of retardation time for the viscous part and relaxation time for the polymer part. In 1982, Giesekus [33] proposed a generalization of Jeffrey's model in which the stress is regularized by an additional nonlinear term. Since then, numerical works of solving viscoelastic fluid flow face the choice of many different constitutive laws that predict the behavior of real viscoelastic fluids under consideration.

A general viscoelastic model can be written as the following

$$\Lambda \stackrel{\triangledown}{\boldsymbol{\sigma}} + h(\boldsymbol{\sigma}, \Lambda)\boldsymbol{\sigma} - 2\eta_p \mathbf{D} + g(\boldsymbol{\sigma}) = 0, \qquad 1.4$$

where Λ and η_p are the relaxation time and the polymer viscosity [46]. As already slightly mentioned, the stress of viscoelastic fluid depends nonlinearly on the deformation rate and is time dependent for the elastic part. Here $\stackrel{\triangledown}{\boldsymbol{\sigma}}$ is the upper-convective derivative, while $h(\boldsymbol{\sigma}, \Lambda)$ and $g(\boldsymbol{\sigma})$ are model-dependant. Viscoelastic models within this family include Oldroyd-B, UCM (Upper convective Maxwell), FENE (Finite extensible nonlinear elastic), PTT (Phan-Thien-Tanner) and Giesekus. In these models Λ and η_p are constants. In the case of Oldroyd-B and UCM, $g(\boldsymbol{\sigma}) = 0$ and $h(\boldsymbol{\sigma}, \Lambda) = 1$, in the case of Giesekus, $h(\boldsymbol{\sigma}, \Lambda) = 1$, while in the case of FENE and PTT, $g(\boldsymbol{\sigma}) = 0$ and $h(\boldsymbol{\sigma}, \Lambda)$ is a nonlinear function.

Among viscoelastic models, Oldroyd-B is the simplest model and utilized very often for viscoelastic benchmark problems. Its generality is well accepted in the CFD community to test different numerical methods and formulations even though this model has an infinite extensional viscosity that leads to an unphysical stress growth in the direction of the extensional flow. Therefore, in this study, it is sufficient to restrict focus on Oldroyd-B model in order to discuss numerics specific for viscoelastic flows. Viscoelastic fluid flow modelling itself is a vast area of research and is not considered in detail in this work. Only the Giesekus model is taken into account as an alternative for comparison.

Since late 1970s, numerical work developed in parallel with development of computer technology. Crochet and Bezy [19] did the first numerical attempt to solve plane flow of viscoelastic

---
[3]Instantaneous elasticity describes a condition of the fluid that memories all past state of stress [44]. The time scale of all memorized stresses is well-known as the relaxation time.

fluid in a contraction, which introduced mixed finite elements within Eulerian framework. In solving the above equation, the convective term $(\mathbf{u} \cdot \nabla)\boldsymbol{\sigma}$ becomes more dominant as the We number increases, and Galerkin formulation starts to have problems. The standard remedy is to apply inconsistent techniques like streamline diffusion or upwind-schemes, see [14] and [51] for implementation. Another way of solving is by using discontinuous Galerkin formulation where the extra stress tensor is discontinuously approximated, see [30]. Both ways of solving try to impose more elliptic character of the resulting discrete system. Introduction of a change of variables came up and it is known as Elastic Viscous Stress Splitting (EVSS), see [8, 9]. Unfortunately, the change of variable does not yield a closed expression. A remedy is to consider $\mathbf{D}$ as a separate variable by an L_2 projection of the velocity gradient. The family of this techniques grows larger with new ideas and today are well-known as DEVSS/-G (Discrete EVSS with $\mathbf{G}$ stemming from an L_2 projection of the transpose of the velocity gradient), see [31, 29]. Another good approach in circumventing the compatibility condition in the case of nonlinearity is the so-called Galerkin/least square (GLS) method, which is introduced by Hughes et. al. [39] for Navier-Stokes equation and extended into Oldroyd-B type by Behr [7]. GLS method provides similar benefits as the above mentioned methods in a way that the resulting set of equations has low order degree, thus it is easier to solve. A rather different approach of reformulating the stress equation was presented by Hulsen et. al. [41], which is based on logarithm of conformation tensor. This method is able to capture the high stress gradient that many suspect as the cause of the numerical breakdown.

Over the years, several issues surrounding viscoelastic simulations become clear. One may summarize these issues to be the reformulation of the model, the choice of finite element pairs and the stabilization techniques. Through these, simulation of viscoelastic flow at high Weissenberg number is not impossible, provided an appropriate model of viscoelastic fluid flow. The first issue deals with the reformulation of the model itself. Since the so-called conformation stress tensor property turns out not being maintained during numerical computation given in the original viscoelastic model, it became necessary to reformulate the equation. Here, reformulation means an introduction of a new variable that may ease the numerical stability, which is described in chapter 2. The second issue is how to discretize the domain in space and time. While discretizing in time is not 'hard', space discretization by finite element functions must preserve some LBB condition as in the original Navier-Stokes problem case. If it is not the case then one remedy is the use of stabilization technique that may ease the choice of the stress function approximation. The whole thing is described in chapter 3.

1.3 Contribution of the thesis

The monolithic treatment of any set of equations preserves highly accurate solutions due to simultaneously considering all numerical variables at every time step, every nonlinear step and every local step of sublinear problems. In fact, a fully coupled set of equations means that the PDE's contains highly nonlinear couplings that leads to physically reasonable solutions. Judging by its ability of preserving accurate solutions, we can say that the monolithic treatment needs two things: i) A strong nonlinear solver which provides information and control of the direction of the global minima in a defect-correction iterative fashion and ii) A robust linear iterative solver which provides fast and accurate solutions.

The main contribution of the thesis is demonstration of the monolithic FEM approach in solving fluid flow model of incompressible Navier-Stokes equation, isothermal or non-isothermal, with non-Newtonian material models, hereby focusing on viscoelastic material model given in LCR. The non-Newtonian material models depend mostly on the nonlinear

1.3. CONTRIBUTION OF THE THESIS

viscosity functions, which can be temperature, shear rate, or even pressure dependent viscosity

$$\begin{cases} \rho\dfrac{\partial \mathbf{u}}{\partial t} + \rho(\mathbf{u} \cdot \nabla)\mathbf{u} = -\nabla p + \nabla \cdot \mathbf{T} \\ \nabla \cdot \mathbf{u} = 0 \end{cases} \quad 1.5$$

with $\mathbf{T} = 2\eta(\dot{\gamma})\mathbf{D}$ and $\dot{\gamma}$ is the shear rate. The monolithic approach can easily, as shown later on Chapter 4, deal with the non-Newtonian fluids given in nonlinear viscosity functions. Yet, the non-Newtonian fluids given in viscoelastic material models behave more different in nature. The so-called relaxation time, which is a material parameter of the viscoelastic fluid, controls the elastic stress-strain relation while fluid is in motion and 'not' in motion. Thus, the total stress of the viscoelastic fluid in motion consists not only of the viscous stress but also of the elastic stress. While viscous stress depends linearly or nonlinearly on the shear rate, elastic stress is simply unknown which depends on the historical path of stress. Hence, the viscoelastic fluid models rely on the elastic stress equation that describes the evolving elastic stress in motion. Consequently, there are additional extra unknowns to be coupled with the incompressible Navier-Stokes equation

$$\begin{cases} \rho\dfrac{\partial \mathbf{u}}{\partial t} + \rho(\mathbf{u} \cdot \nabla)\mathbf{u} = -\nabla p + \eta_s \Delta \mathbf{u} + \nabla \cdot \boldsymbol{\sigma} \\ \nabla \cdot \mathbf{u} = 0 \\ \Lambda \left(\dfrac{\partial \boldsymbol{\sigma}}{\partial t} + (\mathbf{u} \cdot \nabla)\boldsymbol{\sigma} - \nabla \mathbf{u} \cdot \boldsymbol{\sigma} - \boldsymbol{\sigma} \cdot \nabla \mathbf{u}^T\right) + 2\eta_0(\beta - 1)\mathbf{D} + \boldsymbol{\sigma} = 0 \end{cases} \quad 1.6$$

with $\mathbf{T} = 2\eta_s \mathbf{D} + \boldsymbol{\sigma}$. The monolithic treatment of the above equations introduces a new challenge in the choice of the FEM approximation for the new unknowns, which leads to other numerical treatment such as stabilization technique. In the end, this thesis demonstrates that the monolithic treatment can be applied to other reformulation of the stress equation or to other interesting physical problems such as including temperature effect by additional energy equation (later in chapter 6),

$$\begin{cases} \rho\dfrac{\partial \mathbf{u}}{\partial t} + \rho(\mathbf{u} \cdot \nabla)\mathbf{u} = -\nabla p + \eta_s \Delta \mathbf{u} + \dfrac{\eta_p}{\Lambda} \nabla \cdot e^{\boldsymbol{\psi}} + \rho(1 - \gamma\Theta)\mathbf{j} \\ \nabla \cdot \mathbf{u} = 0 \\ \dfrac{\partial \boldsymbol{\psi}}{\partial t} + (\mathbf{u} \cdot \nabla)\boldsymbol{\psi} - (\boldsymbol{\Omega}\boldsymbol{\psi} - \boldsymbol{\psi}\boldsymbol{\Omega}) - 2\mathbf{B} = \dfrac{1}{\Lambda}\left(e^{-\boldsymbol{\psi}} - \mathbf{I}\right) \\ \dfrac{\partial \Theta}{\partial t} + (\nabla\Theta)\mathbf{u} = k_1\,\nabla^2\Theta + k_2\,exp(\boldsymbol{\psi}) : \mathbf{D} \end{cases} \quad 1.7$$

Furthermore in this thesis, the high order finite element pair $Q_2 P_1$ is applied. The use of high order FE approximation is also meant to increase accuracy of the numerical solutions which is shown later in Chapter 5 of benchmarking flow problems.

2

Governing equations

2.1 General Newtonian flow

Any material that flows in continuum with linearly stress dependent on the velocity gradient can be considered as Newtonian flow. One of the main characteristics of this flow is incompressibility. It describes an unconcentrated flow condition on any part of the fluid domain[1]. In other words, what comes in must come out. The other flow characteristic follows Newton's second law that describes the balance of force of a tiny material. The two characteristics, when combined, are well-known as the incompressible Navier-Stokes equation, which describes a Newtonian fluid motion. The reader may find derivation of the equation in the literature [47]. Here, it can be written as

$$\begin{cases} \rho \dfrac{\partial \mathbf{u}}{\partial t} + \rho (\mathbf{u} \cdot \nabla) \mathbf{u} = -\nabla p + \nabla \cdot \mathbf{T} \\ \nabla \cdot \mathbf{u} = 0 \end{cases} \qquad 2.1$$

with the constitutive law of a Newtonian fluid $\mathbf{T} = 2\eta \mathbf{D}$. The above set of equations is already nonlinear and the Newtonian material model gives rise to additional numerical complexity through nonlinear viscosity function such as shear rate or pressure dependent or even both dependent, $\eta(\mathbf{D}, p)$. If it is not nonlinear function, a very small constant value of viscosity produces already bad matrix properties which leads to the well-known high Reynolds number problems, $\mathsf{Re} = \rho \frac{u_c l_c}{\eta}$. By u_c and l_c we refer to the characteristic velocity and length.

The temperature effect of the flow (non-isothermal) is written in term of the Boussinesq approximation. Its coupling with the above Navier-Stokes equations looks like as follows

$$\begin{cases} \rho \dfrac{\partial \mathbf{u}}{\partial t} + \rho (\mathbf{u} \cdot \nabla) \mathbf{u} = -\nabla p + \nabla \cdot \mathbf{T} + \rho (1 - \gamma \Theta) \mathbf{j} \\ \nabla \cdot \mathbf{u} = 0 \\ \dfrac{\partial \Theta}{\partial t} + (\mathbf{u} \cdot \nabla) \Theta = k_1 \nabla^2 \Theta + k_2 \mathbf{D} : \mathbf{D} \end{cases} \qquad 2.2$$

with $\mathbf{j}$ denoting the gravity vector and γ being the thermal expansion. This set of PDE's describes the temperature motion as a transport problem in continuum only. In order to take into account the viscoelastic effects of the flow, one needs additional stress equations that govern the 'polymer' contribution or the elastic part of the fluid. This will be discussed in the following section.

---

[1] By domains we mean open sets.

2.2 Viscoelastic material model

The essence of viscoelastic fluids depends on the ratio between time scales of the fluid (relaxation time Λ) and the flow (retardation time Λ_r), which is later known as the Deborah number. These time scales measure the relative instantaneous elasticity. The first time scale (memory of fluid) denotes the time needed by the stretched material to come back into its relaxed state (by recoiling) irrespective of its initial state, while the second denotes the time scale of the flow without which the fluid behaves like an elastic solid. Another important non dimensional number, the so-called Weissenberg number $\text{We} = \Lambda \frac{u_c}{l_c}$, describes the flow behavior and in some sense also the numerical difficulties. This number is the product of relaxation time of the fluid and the shear rate of the flow. In the following, one may see that the main variable given in viscoelastic models can be described in terms of elastic stress, conformation stress, or even LCR.

2.2.1 Viscoelastic model with conformation stress tensor

The stress equation describing the viscoelastic motion was first proposed by Jeffrey in 1929. Then in 1950, Oldroyd [54] came up with separation of the above mentioned time scales in the model. Giesekus, in 1982, introduced a nonlinear term into the model. Following Oldroyd, it can be written as follows

$$\mathbf{T} + \Lambda \frac{\delta_a \mathbf{T}}{\delta t} = 2\eta_0 \left(\mathbf{D} + \Lambda_r \frac{\delta_a \mathbf{D}}{\delta t} \right) \qquad 2.3$$

where $\mathbf{T}, \mathbf{D} = \frac{1}{2}(\nabla \mathbf{u} + \nabla \mathbf{u}^T), \eta_0, \Lambda, \Lambda_r$ are the extra stress tensor, symmetric velocity gradient, total viscosity, relaxation time, and retardation time respectively. The term $\frac{\delta_a}{\delta t}$ denotes the upper/lower convected material time derivative by setting $a = 1/-1$,

$$\frac{\delta_a \mathbf{T}}{\delta t} = \frac{D\mathbf{T}}{Dt} + \frac{1-a}{2}(\nabla \mathbf{u} \cdot \mathbf{T} + \mathbf{T} \cdot \nabla \mathbf{u}^T) + \frac{1+a}{2}(-\mathbf{T} \cdot \nabla \mathbf{u} - \nabla \mathbf{u}^T \cdot \mathbf{T}), \qquad 2.4$$

which has the nice property of

$$\frac{\delta_a \mathbf{I}}{\delta t} = -2a\mathbf{D}. \qquad 2.5$$

The total viscosity is the well-known zero-shear viscosity, which is the sum of solvent viscosity (η_s) and polymer viscosity (η_p). The extra stress tensor $\mathbf{T}$ consists of two parts namely the viscous stress component, $2\eta_0 \frac{\Lambda_r}{\Lambda} \mathbf{D}$, and the elastic stress component, $\boldsymbol{\sigma}$,

$$\mathbf{T} = 2\eta_0 \frac{\Lambda_r}{\Lambda} \mathbf{D} + \boldsymbol{\sigma}. \qquad 2.6$$

By replacing $\mathbf{T}$ in equation (2.3) with the right hand side of equation (2.6) and setting $a = 1$, we recover the well-known Oldroyd-B model with elastic stress tensor as the main numerical variable (details can be found in Appendix A.1)

$$\Lambda \left(\frac{\partial \boldsymbol{\sigma}}{\partial t} + (\mathbf{u} \cdot \nabla)\boldsymbol{\sigma} - \nabla \mathbf{u} \cdot \boldsymbol{\sigma} - \boldsymbol{\sigma} \cdot \nabla \mathbf{u}^T \right) + 2\eta_0 (\beta - 1)\mathbf{D} + \boldsymbol{\sigma} = 0, \qquad 2.7$$

where $\beta = \frac{\Lambda_r}{\Lambda}$ is the amount of solvent contribution. Next, the conformation stress tensor which has the positive definite property [40] is introduced

$$\boldsymbol{\sigma} = \frac{\eta_p}{\Lambda}(\boldsymbol{\tau} - \mathbf{I}). \qquad 2.8$$

By replacing $\boldsymbol{\sigma}$ in equation (2.7) with $\boldsymbol{\tau}$, we rewrite the Oldroyd-B model in terms of the conformation stress tensor $\boldsymbol{\tau}$ (details can be found in Appendix A.1)

$$\frac{\partial \boldsymbol{\tau}}{\partial t} + \overbrace{(\mathbf{u} \cdot \nabla)\boldsymbol{\tau}}^{\text{convection}} \underbrace{- \nabla \mathbf{u} \cdot \boldsymbol{\tau} - \boldsymbol{\tau} \cdot \nabla \mathbf{u}^T}_{\text{stretching}} + \frac{1}{\Lambda}(\boldsymbol{\tau} - \mathbf{I}) = 0. \qquad 2.9$$

2.2. VISCOELASTIC MATERIAL MODEL

Furthermore, Giesekus model is considered as an alternative in this work which reads

$$\frac{\partial \boldsymbol{\tau}}{\partial t} + \overbrace{(\mathbf{u} \cdot \nabla)\boldsymbol{\tau}}^{\text{convection}} \underbrace{-\nabla \mathbf{u} \cdot \boldsymbol{\tau} - \boldsymbol{\tau} \cdot \nabla \mathbf{u}^T}_{\text{stretching}} + \frac{1}{\Lambda}(\boldsymbol{\tau} - \mathbf{I} + \alpha(\boldsymbol{\tau} - \mathbf{I})^2) = 0. \qquad 2.10$$

We will see later that Giesekus model gives a better mesh converged solution for viscoelastic benchmark and that the limit of We number may be model dependent.

It is worth to note that the above conformation tensor also has an integral form which guarantees the positive definiteness of the conformation tensor and which illustrates the exponential behavior,

$$\boldsymbol{\tau}(t) = \int_{\infty}^{t} \frac{1}{\Lambda} \exp\left(\frac{-(t-s)}{\Lambda}\right) F(s,t) F(s,t)^T ds \qquad 2.11$$

where $F(s,t)$ is the relative deformation gradient [57, 48].

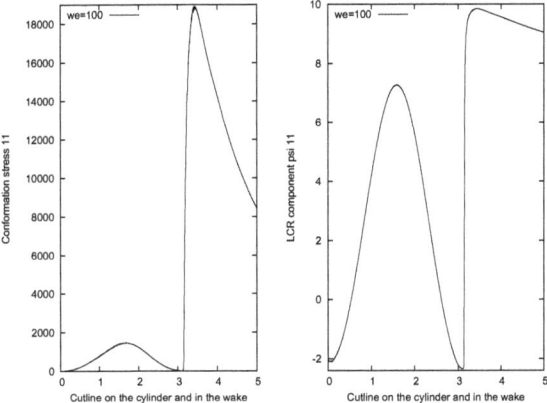

Figure 2.1: Viscoelastic flow around a cylinder for Giesekus model; Cutline of conformation stress tensor τ_{11} (left) and ψ_{11} (right) for We = 100.

The problem in this family of constitutive model is that the normal conformation stress component may go to infinity which limits the possibility of higher We number computation. Hence, having the conformation tensor as the main numerical variable may not be good in such a case. For example, Fig. (2.1) shows a steep gradient of τ_{11} (between 0-20000) in the viscoelastic flow around cylinder for We = 100 with Giesekus model. The viscoelastic model given in LCR is able to capture this effect because the range value of the resulting numerical results is not as dramatic as if we choose equation (2.10), which is between 0-10. So, the situation is far more different if we use LCR, which will be described in the following subsection.

2.2.2 Viscoelastic model with LCR

The positive definite property of the conformation tensor, in equation (2.9), may get lost during numerical computation. Therefore, Fattal and Kupfermann [28] reformulated equation

CHAPTER 2. GOVERNING EQUATIONS

(2.9) by introducing a new logarithmic variable that preserves this property 'by design'. We will shortly revisit how this reformulation is realized: The conformation tensor is replaced by a new variable $\psi = R \, \log(\tau) R^T$ through eigenvalue computation that rotates the τ into its main principle axis (diagonalization process)

$$R^T \tau R = \text{diag}(\lambda_1, \lambda_2) \qquad 2.12$$

with R being an orthogonal matrix. The goal is to decompose the velocity gradient into a symmetric matrix $\mathbf{B}$ that commutes with τ, a pure rotation matrix $\mathbf{\Omega}$ and an additional 'dummy part' matrix $\mathbf{N}\tau^{-1}$ (details can be found in Appendix A.1)

$$\nabla \mathbf{u} = \mathbf{B} + \mathbf{\Omega} + \mathbf{N}\tau^{-1}. \qquad 2.13$$

Here, $\mathbf{N}$ and $\mathbf{\Omega}$ are pure rotational matrices (see [28]). The challenging task is to express those matrices in terms of the velocity gradient. Here, the idea is to take the rotation matrix R and apply it to all components of the velocity gradient. Consequently, we obtain

$$R^T \nabla \mathbf{u} R = R^T \mathbf{B} R + R^T \mathbf{\Omega} R + R^T \mathbf{N} \tau^{-1} R. \qquad 2.14$$

By doing so, we have the possibility to define $\mathbf{B}$ to be commutable with τ and then to define the rest of the matrices $(\mathbf{\Omega}, \mathbf{N})$ in terms of the velocity gradient as shown in [28]. At the end, the matrix $\mathbf{B}$ is not purely extensional, but it is in the same plane as τ.

By inserting the decomposition (2.13) into (2.9), the constitutive law transforms into (2.15) (details can be found in Appendix A.1)

$$\frac{\partial \tau}{\partial t} + (\mathbf{u} \cdot \nabla)\tau - (\mathbf{\Omega}\tau - \tau\mathbf{\Omega}) - 2\mathbf{B}\tau = \frac{1}{\Lambda}(\mathbf{I} - \tau). \qquad 2.15$$

Finally, by replacing the conformation tensor with the new variable $\psi = \log(\tau)$, the Oldroyd-B model evolves to

$$\frac{\partial \psi}{\partial t} + (\mathbf{u} \cdot \nabla)\psi - (\mathbf{\Omega}\psi - \psi\mathbf{\Omega}) - 2\mathbf{B} = \frac{1}{\Lambda}\left(e^{-\psi} - \mathbf{I}\right). \qquad 2.16$$

Consequently, the new set of equations to be solved can be finally rewritten as follows:

$$\begin{cases} \rho\dfrac{\partial \mathbf{u}}{\partial t} + \rho(\mathbf{u} \cdot \nabla)\mathbf{u} = -\nabla p + \eta_s \Delta \mathbf{u} + \dfrac{\eta_p}{\Lambda} \nabla \cdot e^{\psi} \\ \nabla \cdot \mathbf{u} = 0 \\ \dfrac{\partial \psi}{\partial t} + (\mathbf{u} \cdot \nabla)\psi - (\mathbf{\Omega}\psi - \psi\mathbf{\Omega}) - 2\mathbf{B} = \dfrac{1}{\Lambda} f(\psi) \end{cases} \qquad 2.17$$

where

$$f(\psi) = \begin{cases} (e^{-\psi} - \mathbf{I}) & \text{Oldroyd-B} \\ (e^{-\psi} - \mathbf{I}) - \alpha e^{\psi}(e^{-\psi} - \mathbf{I})^2 & \text{Giesekus} \end{cases} \qquad 2.18$$

The source of nonlinearity of equation (2.17) depends highly on the relaxation time scale of the fluid, Λ.

2.3 Boundary and initial conditions

Most of the real flow problems are not physically bounded/closed (i.e. air flow around an airplane). In contrast, the set of PDE's in (2.1, 2.2 or 2.17) must be applied to a "closed" domain by some kind of boundary conditions without which the numerical system can not be

2.3. BOUNDARY AND INITIAL CONDITIONS

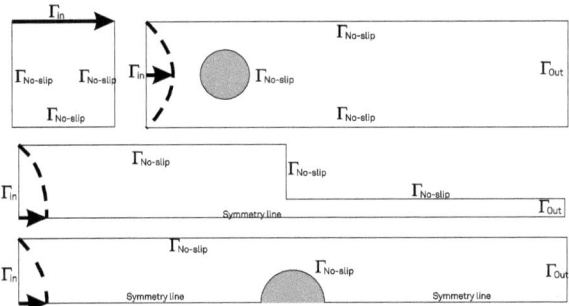

Figure 2.2: Boundary conditions on several configurations.

solved. A small note again, domains are open sets. In the following, we set Γ_{in} and Γ_{out} as the boundary for inflow and outflow, while $\Omega \subset \Re^d$ denotes the computational domain. For several geometrical examples of boundary conditions can be seen in Fig. 2.2.

The usual boundary condition at the inflow is a prescribed velocity profile, $\mathbf{u}(y)$, in two dimensions. At the outflow, the velocity is set to "Do nothing" condition. This natural outflow condition is obtained by integration by parts of the weak form of equation 2.1 which yields in $\eta \partial_n \mathbf{u} - pn = 0$ (see [37]). The inflow boundary profile for the stress variables (or ψ from LCR) must be calculated based on the inflow velocity boundary profile while the outflow stress boundary is set to natural ('Do nothing') boundary conditions [37]. This is necessary because the viscoelastic fluid has a memory. The outflow stress boundary is set to natural boundary conditions. The pressure has no boundary condition since it is an L^2-function. The outflow velocity profile can be set either to Dirichlet (fully developed outflow) or natural boundary condition. We assume the flow to be fully developed at the inflow, and no-slip condition is applied on solid boundaries for the velocity. The no-slip condition requires that the velocity should vanish on rigid boundary walls, $\mathbf{u} = 0$ on $\Gamma_{\text{no-slip}}$. For a given velocity profile, $u_x = 1.5(1 - \frac{y^2}{4})$, the above assumptions generate a stress profile at the inflow [25], which can be written in terms of the conformation stress tensor (details can be found in Appendix A.2)

$$\tau_{yy} = 1, \quad \tau_{xy} = \Lambda \frac{\partial u_x}{\partial y}, \quad \tau_{xx} = 1 + 2\left(\Lambda \frac{\partial u_x}{\partial y}\right)^2. \qquad 2.19$$

The above equations are obtained by the fact that on the inflow the nonlinear terms of equation (2.9), y-velocity component and the x-velocity gradient vanish,

$$(\mathbf{u} \cdot \nabla)\boldsymbol{\tau} = 0, u_y = 0, \partial_x u_x = 0. \qquad 2.20$$

Next, we transform the conformation stress inflow profile into a boundary condition for the new variable ψ in LCR by using their eigenvalues,

$$\lambda_{1,2} = \frac{1}{2}\left[\operatorname{tr}\boldsymbol{\tau} \pm \underbrace{\sqrt{\operatorname{tr}\boldsymbol{\tau}^2 - 4\det\boldsymbol{\tau}}}_{\text{square term}}\right] \qquad 2.21$$

$$\boldsymbol{\psi} = \begin{pmatrix} c & s \\ -s & c \end{pmatrix} \begin{pmatrix} \log\lambda_1 & 0 \\ 0 & \log\lambda_2 \end{pmatrix} \begin{pmatrix} c & -s \\ s & c \end{pmatrix} \qquad 2.22$$

CHAPTER 2. GOVERNING EQUATIONS

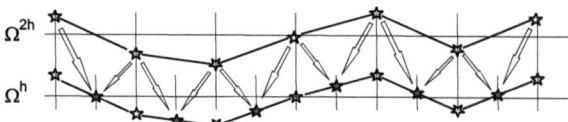

Figure 2.3: Interpolations from coarse to fine grid.

with the property $c^2 + s^2 = 1$. In practice, it is helpful to always check that the square term in equation (2.21) exists. Initial conditions are zero vectors in most cases, $\mathbf{u}(x, y, t) = 0$. But in some cases, initial solutions can be some developed solutions from lower level solutions $\mathbf{u}(x, y)^h = \mathbf{I}_h^{2h} \mathbf{u}(x, y)^{2h}$ or from solutions of lower parameter values (Re or We). If the first case is used then the initial solution will be interpolated first into one level higher. The interpolation can be visualized in one dimension by Fig. 2.3. In the non steady calculation, initial solutions can be some developed solutions from previous time step, $\mathbf{u}(x, y, t^n) = \mathbf{u}(x, y, t^{n-1})$.

3

The discrete time and space system

In solving the full system of equation (2.17) we will discretize it first in time with a fully implicit second order time integrator such as Crank-Nicolson or fractional step ϑ-scheme and in space with FEM. A monolithic way of solving the discrete system allows a proportional numerical treatment between a direct steady and the corresponding nonstationary approach, here by adding a scaled mass matrix with the time step size for the latter approach. In this way, a bigger time step size can be applied and the time step size dependency of mesh size can be avoided.

In the case of space discretization, quadrilateral elements are used for the subdivision of the domain. The mapping between reference and real elements is then formulated on this element. Regular and local refinement are also incorporated within these elements, too. The important point in this chapter is the choice of the finite element space function which should satisfy the LBB condition. If this condition can not be satisfied, then an appropriate stabilization technique is a must. In the case of the edge-oriented stabilization technique, the standard FEM stencil needs to be extended for the implementation.

3.1 The fully implicit time discretization

We apply implicit 2nd order time stepping method to preserve the high accuracy and robustness in nonstationary flow simulations, for instance the Crank-Nicolson or Fractional-Step-θ scheme. This implicit scheme allows adaptive time stepping due to accuracy reasons only [68] and does not depend on CFL-like restrictions. This method is very suitable for physically time dependent problems such as MIT benchmark 2001[17], von Karman vortex shedding in a flow around cylinder[73]. By setting the density to be uniform with value of $\rho = 1$, then the set of equations (2.17) is discretized in time as follows: Given $\mathbf{u}^n, \boldsymbol{\psi}^n$ and $\Delta t = t_{n+1} - t_n$, we seek solutions for the next time step $\mathbf{u}, p, \boldsymbol{\psi}$

$$\frac{\mathbf{u} - \mathbf{u}^n}{\Delta t} +$$

$$\theta \left[\mathbf{u} \cdot \nabla \mathbf{u} - (\eta_s \Delta \mathbf{u} + \frac{\eta_p}{\Lambda} \nabla \cdot e^{\boldsymbol{\psi}}) \right] + \nabla p \qquad 3.1$$

$$+ (1 - \theta) \left[\mathbf{u}^n \cdot \nabla \mathbf{u}^n - (\eta_s \Delta \mathbf{u}^n + \frac{\eta_p}{\Lambda} \nabla \cdot e^{\boldsymbol{\psi}^n}) \right] = 0$$

$$\nabla \cdot \mathbf{u} = 0 \qquad 3.2$$

where $\mathbf{u}^n \sim \mathbf{u}(t_n)$. As one can see, the pressure space is discretized fully implicitly. The LCR equation in the Oldroyd-B model is discretized in the same way so that

$$\frac{\boldsymbol{\psi} - \boldsymbol{\psi}^n}{\Delta t} + \qquad 3.3$$

CHAPTER 3. THE DISCRETE TIME AND SPACE SYSTEM

$$\theta \left[\mathbf{u} \cdot \nabla \psi - (\mathbf{\Omega}(\mathbf{u}).\psi - \psi.\mathbf{\Omega}(\mathbf{u})) - 2\mathbf{B}(\mathbf{u}) - \frac{1}{\Lambda}\left(e^{-\psi} - \mathbf{I}\right) \right]$$

$$+ (1-\theta) \left[\mathbf{u}^n \cdot \nabla \psi^n - (\mathbf{\Omega}(\mathbf{u}^n).\psi^n - \psi^n.\mathbf{\Omega}(\mathbf{u}^n)) - 2\mathbf{B}(\mathbf{u}^n) - \frac{1}{\Lambda}\left(e^{-\psi^n} - \mathbf{I}\right) \right] = 0.$$

The divergence of the exponential operator in equation (3.1) is approximated by the divergence of the conformation tensor via eigenvalue decomposition, which is explained in the previous section as part of LCR. So, the equations above involve solving a typical nonlinear saddle point problem. Given $\mathbf{u}^n, \psi^n$ and $\Delta t = t_{n+1} - t_n$, we seek solutions for the next time step $\mathbf{u}, p, \psi$

$$\mathbf{u} + \Delta t\,\theta\left[A_\mathbf{u}(\mathbf{u})\mathbf{u} + C_{\exp}\psi\right] + \Delta t B p = \mathbf{u}^n - \Delta t(1-\theta)\left[A_\mathbf{u}(\mathbf{u}^n)\mathbf{u}^n + C_{\exp}\psi^n\right]$$

$$B^T \mathbf{u} = 0 \qquad 3.4$$

$$\psi + \Delta t\,\theta\left[A_\psi(\mathbf{u})\psi + F_\nabla(\mathbf{u})\right] = \psi^n - \Delta t(1-\theta)\left[A_\psi(\mathbf{u}^n)\psi^n + F_\nabla(\mathbf{u}^n)\right]$$

with operators

$$A_\psi(\mathbf{u})\psi = N(\mathbf{u})\psi + G_\nabla(\mathbf{u})\psi + H_{\exp^{-1}}\psi \qquad 3.5$$

$$A_\mathbf{u}(\mathbf{u})\mathbf{u} = N(\mathbf{u})\mathbf{u} + \eta_s L(\mathbf{u})\mathbf{u} \qquad 3.6$$

$$N(\mathbf{u})\mathbf{u} = \mathbf{u} \cdot \nabla \mathbf{u} \qquad 3.7$$

$$N(\mathbf{u})\psi = \mathbf{u} \cdot \nabla \psi \qquad 3.8$$

$$L(\mathbf{u})\mathbf{u} = -\Delta \mathbf{u} \qquad 3.9$$

$$C_{\exp}\psi = -\frac{\eta_p}{\Lambda}\nabla \cdot e^\psi \qquad 3.10$$

$$G_\nabla(\mathbf{u})\psi = -(\mathbf{\Omega}(\mathbf{u}).\psi - \psi.\mathbf{\Omega}(\mathbf{u})) \qquad 3.11$$

$$H_{\exp^{-1}}\psi = -\frac{1}{\Lambda}(e^{-\psi} - \mathbf{I}) \qquad 3.12$$

$$F_\nabla(\mathbf{u}) = -2\mathbf{B}(\mathbf{u}) \qquad 3.13$$

By introducing a mass matrix operator, M, and other nonlinear operators $S_\mathbf{u}\mathbf{u} = [M + \Delta t\,\theta A_\mathbf{u}(\mathbf{u})]\mathbf{u}$ and $S_\psi\psi = [M + \Delta t\,\theta A_\psi(\mathbf{u})]\psi$, we can rewrite the set of equations above as follows: Given $\mathbf{u}^n, \psi^n$ and $\Delta t = t_{n+1} - t_n$, we seek solutions for the next time step $\mathbf{u}, p, \psi$

$$S_\mathbf{u}\mathbf{u} + \Delta t\,\theta C_{\exp}\psi + \Delta t B p = \text{rhs } \mathbf{u} \qquad 3.14$$

$$B^T \mathbf{u} = 0 \qquad 3.15$$

$$S_\psi\psi + \Delta t\,\theta F_\nabla(\mathbf{u}) = \text{rhs } \psi \qquad 3.16$$

with

$$\text{rhs } \mathbf{u} = \mathbf{u}^n - \Delta t(1-\theta)\left[A_\mathbf{u}(\mathbf{u}^n)\mathbf{u}^n + C_{\exp}\psi^n\right] \qquad 3.17$$

$$\text{rhs } \psi = \psi^n - \Delta t(1-\theta)\left[A_\psi(\mathbf{u}^n)\psi^n + F_\nabla(\mathbf{u}^n)\right]. \qquad 3.18$$

By choosing $\theta = \frac{1}{2}$, we obtain the fully implicit Crank-Nicolson method with second order accuracy.

3.2 The Finite Element Method

Now that we have a time discretized equation, we discretize the space by means of FEM. The finite element is well-known for its flexibility in dealing with any complex PDE's and

3.2. THE FINITE ELEMENT METHOD

geometry. Hence, it becomes favorable for many industrial purposes. It works in a weak formulation which means the strong formulation is tested with an arbitrary function, which is called a test function, see [6], and integrated over the domain. Now, let's consider equation (2.17) by introducing 3 test functions (v, φ, χ), which are not defined yet,

$$\begin{cases} \int_T \int_\Omega \left(\rho \frac{\partial \mathbf{u}}{\partial t} + \rho(\mathbf{u} \cdot \nabla)\mathbf{u} + \nabla p - \eta_s \Delta \mathbf{u} - \frac{\eta_p}{\Lambda} \nabla \cdot e^{\boldsymbol{\psi}} \right) v \, dV dt = 0 \\ \int_T \int_\Omega (\nabla \cdot \mathbf{u}) \, \varphi \, dV dt = 0 \\ \int_T \int_\Omega \left(\frac{\partial \boldsymbol{\psi}}{\partial t} + (\mathbf{u} \cdot \nabla)\boldsymbol{\psi} - (\boldsymbol{\Omega}\boldsymbol{\psi} - \boldsymbol{\psi}\boldsymbol{\Omega}) - 2\mathbf{B} - \frac{1}{\Lambda} f(\boldsymbol{\psi}) \right) \chi \, dV dt = 0 \end{cases} \quad 3.19$$

The above finite element integrations are simplified by integrations by part for the second order term and pressure part,

$$\int_T \int_\Omega -\eta_s \Delta \mathbf{u} \, v \, dV dt = \int_T \int_\Omega \eta_s \nabla \mathbf{u} \, \nabla v \, dV dt$$

$$\text{and} \quad \int_T \int_\Omega \nabla p \, v \, dV dt = -\int_T \int_\Omega p \, \nabla v \, dV dt \quad 3.20$$

which requires that $\mathbf{u}$ be only a 'H_0^1 function'[1]. Now, we can write the (simplified) weak formulation

$$\begin{cases} \int_T \int_\Omega \left(\rho \frac{\partial \mathbf{u}}{\partial t} + \rho(\mathbf{u} \cdot \nabla)\mathbf{u} - \frac{\eta_p}{\Lambda} \nabla \cdot e^{\boldsymbol{\psi}} \right) v \, dV dt \\ \qquad + \int_T \int_\Omega (-p\mathbf{I} + \eta_s \nabla \mathbf{u}) \, \nabla v \, dV dt = 0 \\ \int_T \int_\Omega (\nabla \cdot \mathbf{u}) \, \varphi \, dV dt = 0 \\ \int_T \int_\Omega \left(\frac{\partial \boldsymbol{\psi}}{\partial t} + (\mathbf{u} \cdot \nabla)\boldsymbol{\psi} - (\boldsymbol{\Omega}\boldsymbol{\psi} - \boldsymbol{\psi}\boldsymbol{\Omega}) - 2\mathbf{B} - \frac{1}{\Lambda} f(\boldsymbol{\psi}) \right) \chi \, dV dt = 0 \end{cases} \quad 3.21$$

This simplification, later on, allows a larger class of solutions than the class of solutions for the corresponding strong formulation. Up to this point, we have not yet introduced the finite element expansion of the discrete solutions for $\mathbf{u}^h, p^h, \boldsymbol{\psi}^h$, which is fundamental in finite element,

$$\mathbf{u}^h = \sum_{j=1}^{NN_\mathbf{u}} \mathbf{u}_j \, v_j, \quad p^h = \sum_{i=1}^{NN_p} p_j \, \varphi_j, \quad \boldsymbol{\psi}^h = \sum_{i=1}^{NN_{\boldsymbol{\psi}}} \boldsymbol{\psi}_j \, \chi_j \quad 3.22$$

where NN is number of unknowns for each variable. The total number of unknowns leads to the total degree of freedoms on each mesh level. Here, $\mathbf{u}_j, p_j, \boldsymbol{\psi}_j$ are the nodal values and v_j, φ_j, χ_j are the so-called basis function. Throughout this study test function and basis function are taken to be the same. Hence, the discretized weak formulation may be written the same as equation (3.21) but with extra superscript 'h' which denotes the discrete finite element solutions.

Now, the most difficult part is to choose what kind of test/basis function that may fit well into the above problems. In particular, we are dealing with a mixed test/basis function between velocity, pressure and stress numerical variable. The mixed formulation is subject to a well-known condition, the so-called LBB condition, see [34]. The main issues in solving the already mentioned set of equations are not only the robustness and efficiency but also

---

[1] H_0^1 is a Hilbert space which vanishes at the boundary, see [13]

CHAPTER 3. THE DISCRETE TIME AND SPACE SYSTEM

a reliable accuracy of the numerical solution. For these reasons, we utilize the LBB-stable conforming finite element pair Q_2P_1, Fig. 3.2. It is known to be one of the "best" Stokes elements (see [2], [38], the contributions according to [38] and in the proceeding [6]), that means most accurate and robust finite element pairs for highly viscous incompressible flow, particularly together with local grid refinement techniques via hanging nodes.

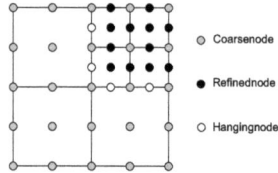

Figure 3.1: Degrees of freedom in locally refined element.

Local refinement technique is meant to reduce the global degrees of freedom. The use of hanging nodes is in a proper way such that the values at hanging nodes must satisfy the continuity constraint of the neighboring nodes. In this way, the finite element test/basis function remains globally continuous and hence conforming, (see [15, 52, 74]). The refinement is done a priori, i.e. without any error indicator.

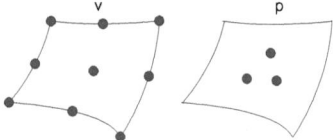

Figure 3.2: The Q_2P_1 pair.

3.3 The choice of finite element space

The choice of the FEM spaces for the (Navier-)Stokes problem is subject to the well-known compatibility condition between the velocity and pressure spaces, the so-called inf − sup condition given in discrete version (LBB) [34]

$$\sup_{\mathbf{u}\in\mathbf{V}_h} \frac{\int_\Omega \mathrm{div}\,\mathbf{u}\,q\,dx}{\|\mathbf{u}\|_{1,\Omega}} \geq \alpha \|q\|_{0,\Omega} \quad \text{for all } q \in Q_h. \qquad 3.23$$

The chosen finite element pair above is compatible according to equation (3.23). Similarly, adding the weak form of the constitutive equation for $\boldsymbol{\sigma}$ imposes further compatibility constraints onto the choice of the approximations spaces for the triple velocity-pressure-stress [13],

$$\sup_{\boldsymbol{\sigma}\in\mathbf{W}_h} \frac{\int_\Omega \mathrm{div}\,\boldsymbol{\sigma}\,\mathbf{u}\,dx}{\|\boldsymbol{\sigma}\|_{0,\Omega}} \geq \gamma \|\mathbf{u}\|_{1,\Omega} \quad \text{for all } \mathbf{u} \in \mathbf{v}_h, \qquad 3.24$$

where α and γ are two mesh-independent constants, $\|\cdot\|_{1,\Omega}$ and $\|\cdot\|_{0,\Omega}$ are the standard $H^1(\Omega)$ and $L^2(\Omega)$ norms and $\mathbf{V}_h \times Q_h \times \mathbf{W}_h \subset (H^1(\Omega))^2 \times L_0^2(\Omega) \times (L^2(\Omega))^4$. Element pairs that satisfy equation (3.24) are also said to be compatible. Fortin and Pierre [32] have shown that

3.3. THE CHOICE OF FINITE ELEMENT SPACE

in the absence of the viscous contribution, the (standard) discrete spaces have to satisfy the following conditions:

1. The velocity-pressure spaces must be compatible with respect to equation (3.23).

2. If the elastic stress tensor $\boldsymbol{\sigma}$ is approximated by a discontinuous FEM space, the deformation tensor must be a member of the same space

$$\mathbf{D} = \frac{1}{2}(\nabla \mathbf{u} + \nabla \mathbf{u}^T) \in \mathbf{W}_h, \quad \text{for all} \quad \mathbf{u} \in \mathbf{V}_h. \qquad 3.25$$

3. If the same tensor is approximated by a continuous FEM space, the number of local degrees of freedom must be larger than that used for the velocity space.

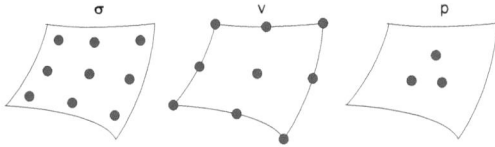

Figure 3.3: Discontinuous quadratic stress; bi-quadratic velocity; discontinuous linear pressure element pair of Fortin and Fortin.

The first requirement is quite obvious from equation (3.23) and is not an issue at this point. Yet, the last two requirements evolve from the need to satisfy equation (3.24) which is a real challenge on mixed finite element for viscoelastic problem. Fortin and Fortin [30] have shown that the second requirement can be fulfilled by using a discontinuous Galerkin approach, see Fig. 3.3, while, in [51], Marchal and Crochet have proposed a subcell discretization to enrich the local degrees of freedom for the stress space, thus fulfill the third requirement.

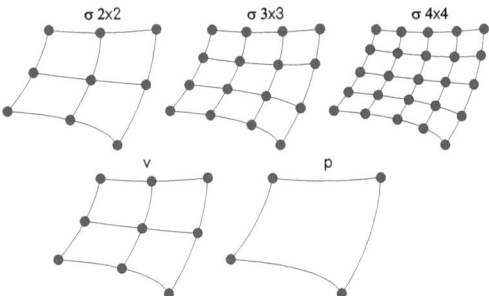

Figure 3.4: Bi-linear stress sub-elements; bi-quadratic velocity; bi-linear pressure element pair of Marchal and Crochet.

In contrast to the requirements above, Baranger and Sandri [5] have shown that a three fields formulation of Stokes's problem and Oldroyd model does not need the equation (3.24) to be fulfilled, thus allows a much larger class of discretization schemes. Further review of mixed finite element method can be found in the work of Baaijens [4, 3] and Fortin et al. [29] which leads to the DEVSS (Discrete Elastic Viscous Splitting). So, this study restricts the

CHAPTER 3. THE DISCRETE TIME AND SPACE SYSTEM

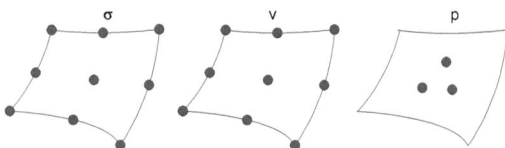

Figure 3.5: Bi-quadratic velocity/stress; discontinuous linear pressure element pair.

finite element pair to the triple $Q_2/P_1^{\text{disc}}/Q_2$, see Fig. 3.5, which according to Baranger and Sandri [5] is not necessarily affected by the last 2 conditions in the case of Oldroyd-B and Giesekus model. Yet, in order to ensure that the element pair is stable, a jump stabilization technique is introduced in the last sections of this chapter. So, the interpolation functions are given as:

$$\mathbf{V}_h = \{\mathbf{v}_h \in (H_0^1(\Omega_h))^2, \quad \mathbf{v}_{h|T} \in (Q_2(T))^2 \;\; \forall T \in \mathcal{T}_h, \quad \mathbf{v}_h = 0 \;\; \text{on} \;\; \partial\Omega_h\},$$

$$Q_h = \{p_h \in L^2(\Omega_h), \quad p_{h|T} \in P_1(T) \;\; \forall T \in \mathcal{T}_h\},$$

$$\mathbf{W}_h = \{\boldsymbol{\sigma}_h \in (L^2(\Omega_h))^4, \quad \boldsymbol{\sigma}_{h|T} \in (Q_2(T))^4 \;\; \forall T \in \mathcal{T}_h\}$$

by considering for each $T \in \mathcal{T}_h$ the bilinear transformation $\psi_T : \hat{T} \to T$ onto the unit square T. So, $Q_2(T)$ is defined by

$$Q_2(T) = \left\{ q \circ \psi_T^{-1} : q \in \text{span} < 1, x, y, xy, x^2, y^2, x^2y, y^2x, x^2y^2 > \right\} \quad 3.26$$

with nine local degrees of freedom located at the vertices, midpoints of the edges and in the center of the quadrilateral. The space $P_1(T)$ consists of linear functions defined by

$$P_1(T) = \left\{ q \circ \psi_T^{-1} : q \in \text{span} < 1, x, y > \right\} \quad 3.27$$

with the function value and both partial derivatives, located in the center of the quadrilateral, as its three local degrees of freedom. The velocity-pressure inf-sup condition is satisfied (see [10]) as well as the velocity-stress inf-sup condition in the presence of a purely viscous contribution[5]. However, the combination of the bilinear transformation ψ with a linear function on the reference square $P_1(\hat{T})$ would imply that the basis on the reference square does not contain the full bilinear basis. So, the method can be only first order accurate on general meshes (see [2, 10])

$$\|p - p_h\|_0 = O(h). \quad 3.28$$

The standard remedy (see [2, 60, 69]) is to consider a local coordinate system (ξ, η) obtained by joining the midpoints of the opposing faces of T. Then, we set on each element T

$$P_1(T) := \text{span} < 1, \xi, \eta > . \quad 3.29$$

In this case, the inf-sup condition is also satisfied and the second order approximation is recovered for the pressure as well as for the velocity gradient (see[10, 34]),

$$\|p - p_h\|_0 = O(h^2) \quad \text{and} \quad \|\nabla u - \nabla u_h\|_0 = O(h^2). \quad 3.30$$

For a smooth solution, the approximation error for the velocity in the L_2-norm is of order $O(h^3)$ which can easily be demonstrated for prescribed polynomials or for smooth data on appropriate domains [10].

In the following, we set up the total discrete nonlinear operator based on the chosen finite element space. The above mentioned mixed FEM discretization together with appropriate initial and boundary conditions results in a set of discrete nonlinear equations which may be written as follows

$$\underbrace{\begin{pmatrix} S_{\mathbf{u}}(\mathbf{u}) & \Delta t\, \theta C_{\exp} & \Delta t B \\ \Delta t\, \theta F_\nabla(\mathbf{u}) & S_\psi(\mathbf{u}) & 0 \\ B^T & 0 & 0 \end{pmatrix}}_{\text{Nonlinear operator } \mathcal{A}} \begin{pmatrix} \mathbf{u} \\ \psi \\ p \end{pmatrix} = \begin{pmatrix} \text{rhs } \mathbf{u} \\ \text{rhs } \psi \\ \text{rhs } p \end{pmatrix} \qquad 3.31$$

with operators that are already described in the last subsection. Here, $S_{\mathbf{u}}(\mathbf{u})$ consists of the mass, diffusive and convective operators, while $S_\psi(\mathbf{u})$ consists of the mass, transformation, convection and rotation operators. It is visible from equations (3.14-3.18) that the mass matrix and the time dependent right hand side cancel out in the case of steady calculation, thus the nonlinear operator may be written as follows:

$$\underbrace{\begin{pmatrix} S_{\mathbf{u}}(\mathbf{u}) & C_{\exp} & B \\ F_\nabla(\mathbf{u}) & S_\psi(\mathbf{u}) & 0 \\ B^T & 0 & 0 \end{pmatrix}}_{\text{Nonlinear operator } \mathcal{A}} \begin{pmatrix} \mathbf{u} \\ \psi \\ p \end{pmatrix} = \begin{pmatrix} \text{rhs } \mathbf{u} \\ \text{rhs } \psi \\ \text{rhs } p \end{pmatrix} \qquad 3.32$$

This typical saddle point problem contains highly nonlinear couplings between the Navier-Stokes and the LCR equation, which can be seen from the operator $C_{\exp}$ and $F_\nabla(\mathbf{u})$. While the operator $S_{\mathbf{u}}(\mathbf{u})$ and $S_\psi(\mathbf{u})$ are already nonlinear, the full nonlinear system becomes very hard to solve in the case of direct steady solution. In the non steady case, equation (3.31), the nonlinearity is somehow 'balanced' by the chosen time step, i.e. small time step size would ease the two coupling operators $C_{\exp}$, $F_\nabla(\mathbf{u})$ contribution as well as the two nonlinear operators contribution, $S_{\mathbf{u}}(\mathbf{u})\mathbf{u} = [M + \Delta t\, \theta A_{\mathbf{u}}(\mathbf{u})]\mathbf{u}$ and $S_\psi(\mathbf{u})\psi = [M + \Delta t\, \theta A_\psi(\mathbf{u})]\psi$. Yet, in the direct steady case, there are two quantities: Re and We numbers which make this highly nonlinear operator, $\mathcal{A}$, even more difficult to solve. Thus, the quality of the resulting discrete solutions may lead to non-physical oscillations. So, measurement towards minimizing the oscillation effects is by introducing a term that stabilize the discrete systems. This will be explained in the following section.

3.4 Artificial diffusion stabilization

The total nonlinear system of equation (3.32) mainly comes from the operator $S_\psi(\mathbf{u})$ that has a convective term of $(\mathbf{u} \cdot \nabla)\psi$. Unlike operator $S_{\mathbf{u}}(\mathbf{u})$ that has a linear operator $L(\mathbf{u})$ inside, see equation (3.6), operator $S_\psi \psi$ does not have one. So, why not give a small amount of artificial diffusion, in the form of a linear operator of $\Delta \psi$, that may help the numerical stability as the following

$$\underbrace{\begin{pmatrix} S_{\mathbf{u}}(\mathbf{u}) & C_{\exp} & B \\ F_\nabla(\mathbf{u}) & S_\psi(\mathbf{u}) + L_\psi(\psi) & 0 \\ B^T & 0 & 0 \end{pmatrix}}_{\text{Nonlinear operator } \mathcal{A}} \begin{pmatrix} \mathbf{u} \\ \psi \\ p \end{pmatrix} = \begin{pmatrix} \text{rhs } \mathbf{u} \\ \text{rhs } \psi \\ \text{rhs } p \end{pmatrix} \qquad 3.33$$

CHAPTER 3. THE DISCRETE TIME AND SPACE SYSTEM

In practice, this contribution should be controlled by a parameter $h\gamma$ which would decrease with mesh refinement, $h \to 0$. In this simple way, a compromise would be found between accuracy and stability of the numerical computation. A more robust stabilization is still needed which is described in the following section.

3.5 EO-FEM stabilization

Bonito and Burman [12] have shown that inf-sup stability as well as stability for convection-dominated flows can be obtained by adding a consistent stabilization term penalizing the jump of the solution gradient over element edges E (with h_E denoting the length of the edge). This term can be written in the following form, see also [72, 71] for more details:

$$J_u = \sum_{\text{edge E}} \max(\gamma\eta h_E, \gamma^* h_E^2) \int_E [\nabla \mathbf{u}] : [\nabla \mathbf{v}] ds \qquad 3.34$$

The same technique can be utilized in the equations for the stress tensor (or for the logarithm of the conformation stress in the case of LCR), particularly for relaxing the choice of the stress space in the absence of a pure viscous contribution. Moreover, since the convective terms of the constitutive equations also require appropriate stabilization techniques, the corresponding edge-oriented jump terms for the stress have been introduced which read as follows

$$J_\sigma = \sum_{\text{edge E}} \gamma h_E^2 \int_E [\nabla \boldsymbol{\sigma}] : [\nabla \chi] ds \qquad 3.35$$

which will be added into the operator $S_{\boldsymbol{\psi}}(\mathbf{u})$,

$$\underbrace{\begin{pmatrix} S_{\mathbf{u}}(\mathbf{u}) & C_{\text{exp}} & B \\ F_\nabla(\mathbf{u}) & S_{\boldsymbol{\psi}}(\mathbf{u}) + J_\sigma & 0 \\ B^T & 0 & 0 \end{pmatrix}}_{\text{Nonlinear operator } \mathcal{A}} \begin{pmatrix} \mathbf{u} \\ \psi \\ p \end{pmatrix} = \begin{pmatrix} \text{rhs } \mathbf{u} \\ \text{rhs } \psi \\ \text{rhs } p \end{pmatrix} \qquad 3.36$$

For simplicity we call this jump stabilization technique (J_σ) as EO-FEM. Here, the gradient of the trial χ_i and test χ_j functions are of the same polynomial which belongs to Q_2,

$$J_\sigma = \sum_{\text{edge E}} \gamma h_E^2 \int_E \boldsymbol{\sigma}_h [\nabla \chi_i] : [\nabla \chi_j] ds \qquad 3.37$$

where the jump term $[\nabla \chi] = \nabla \chi^+ - \nabla \chi^-$ denotes the difference of the gradient values (it can be of any other type of values, see Ref. [55]) between two adjacent elements that share the same edge E. Here, a (constant) parameter γ is provided which is independent from the physical problem, and h_E, as already mentioned, is the length of the edge under consideration which guarantees the weakly consistency of the solution by refinement, see [12]. Equation (3.37) expands to

$$J_\sigma = \sum_{\text{edge E}} \gamma h_E^2 \int_E \boldsymbol{\sigma}_h (\nabla \chi_i^+ - \nabla \chi_i^-)(\nabla \chi_j^+ - \nabla \chi_j^-) ds$$

$$J_\sigma = \sum_{\text{edge E}} \gamma h_E^2 \underbrace{\int_E \boldsymbol{\sigma}_h \nabla \chi_i^+ (\nabla \chi_j^+ - \nabla \chi_j^-) ds}_{\text{first}} - \underbrace{\int_E \boldsymbol{\sigma}_h \nabla \chi_i^- (\nabla \chi_j^+ - \nabla \chi_j^-) ds}_{\text{second}} \qquad 3.38$$

3.5. EO-FEM STABILIZATION

The integration of the above equation can be performed by looping over all edges in the domain and then calling the trial/test functions from the two adjacent elements, see Ref. [55]. In this study, the looping remains over all elements. Thus, each internal edge will be visited twice. In this way, equation (3.38) is completed after the first and second visit.

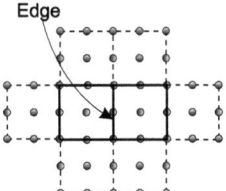

Figure 3.6: Extension of local degrees of freedom

In order to treat the jump term appropriately within the standard FEM discretization, the extension of the standard FEM stencil is needed since there are no sufficient places to put extra non-zero entries in the standard FEM stencil. Fig. 3.6 visualizes the increasing local support of non-zero entries in 2D problem. This is due to the fact that equation (3.38) involves 2 adjacent elements that increase the local degrees of freedom from originally 9 (one element) to 15 (2 elements). The non-stabilized support can be considered as the first stage of nonzero entries implemented in standard FEM, thus the stabilized support needs the second stage of nonzero entries.

4
The Numerical Solver

This chapter briefly describes the solution methods for the discrete system that has been built by the finite element discretization in the previous chapter. The total discrete system in the case of viscoelastic fluid flow is of mixed elliptic-hyperbolic type in the direct steady approach. The task is to solve a set of nonlinear equations given in equation (4.1),

$$\underbrace{\begin{pmatrix} S_\mathbf{u}(\mathbf{u}) & C_{\exp} & B \\ F_\nabla(\mathbf{u}) & S_\psi(\mathbf{u}) & 0 \\ B^T & 0 & 0 \end{pmatrix}}_{\text{Nonlinear operator } \mathcal{A}} \begin{pmatrix} \mathbf{u} \\ \psi \\ p \end{pmatrix} = \begin{pmatrix} \text{rhs } \mathbf{u} \\ \text{rhs } \psi \\ \text{rhs } p \end{pmatrix} \qquad 4.1$$

which can be written as $\mathcal{A}\mathbf{x} = \mathbf{b}$.

4.1 Outer Newton solver

The task of the nonlinear solver becomes very crucial with respect to the numerical stability. There are only few standard choices in this case, namely: fixed point iteration and Newton iteration. Both methods are well-known and widely used in numerical computations. In the general case, the Newton iteration is always preferable since it gives a faster convergence rate than fixed point iteration. But this can be valid only if the initial solution is close enough to the final solution. Thus, Newton iteration should be taken with care. In the case of the well-known Stokes or Navier-Stokes problem for small Re numbers, Newton method converges quadratically. But for high Re numbers, where the solutions may not be smooth due to bad matrix properties, Newton method should be damped appropriately. A careful damping factor is chosen by a line search method which is one of the root finding techniques for nonlinear systems of equations [24, 59]. This guarantees the stability of the Newton iteration. Within

CHAPTER 4. THE NUMERICAL SOLVER

a direct steady, one Newton step can be described as follows:

Algorithm 1: One step Newton

Input: nonlinear parameters, i.e. tolerance
1. set $l = 0$
2. Construct the residual vector $\mathbf{r}(\mathbf{x}^l) = \mathcal{A}\mathbf{x}^l - \mathbf{b}$
3. Construct the Jacobian $\mathbf{J}(\mathbf{x}^l) = \frac{\partial \mathbf{r}}{\partial \mathbf{x}}(\mathbf{x}^l)$
4. Solve the linear system for the correction $\mathbf{J}(\mathbf{x}^l)\delta\mathbf{x}^l = \mathbf{r}(\mathbf{x}^l)$
5. Compute the optimal damping factor $\omega^l \in (-1, 0]$
6. Find a better approximation $\mathbf{x}^{l+1} = \mathbf{x}^l - \omega^l \delta\mathbf{x}^l$
7. **if** *the norm of residual vector $\mathbf{r}(\mathbf{x}^{l+1})$ is below a certain threshold* **then**
   | $l = l + 1$
   | back to step 2
8. **end**

4.2 The Jacobian matrix

The Newton iteration evidently needs to compute the first derivative of the residual with respect to the current solution vector (sometimes it is called tangent stiffness matrix) at every Newton step and level (The Jacobian: step 3 from the above algorithm).

4.2.1 Analytic Jacobian

The traditional way to achieve this is by hand-coding of an analytical expressions by means of taking the Fréchet derivative of the nonlinear operator $\mathcal{A}$. We start from a residual vector $\mathbf{r}(\mathbf{x}^l) = \mathcal{A}\mathbf{x} - \mathbf{b} = 0$ and expand it for all numerical components $(u_1, u_2, p, \psi_{11}, \psi_{12}, \psi_{22})$,

$$r_i(u_1) = \int_\Omega \frac{u_1 - u_1^n}{\Delta t} v_i + \frac{1}{2}\int_\Omega (u_1\partial_1 u_1 + u_2\partial_2 u_1)\, v_i - \int_\Omega p\, \partial_1 v_i$$
$$+ \frac{1}{2}\int_\Omega \eta_s \nabla u_1 \cdot \nabla v_i - \frac{1}{2}\int_\Omega \frac{\eta_p}{\Lambda}\nabla \cdot e^\psi\, v_i - \frac{1}{2}\int_\Omega f^{u_1} v_i \qquad 4.2$$

$$r_i(u_2) = \int_\Omega \frac{u_2 - u_2^n}{\Delta t} v_i + \frac{1}{2}\int_\Omega (u_1\partial_1 u_2 + u_2\partial_2 u_1)\, v_i - \int_\Omega p\, \partial_2 v_i$$
$$+ \frac{1}{2}\int_\Omega \eta_s \nabla u_2 \cdot \nabla v_i - \frac{1}{2}\int_\Omega \frac{\eta_p}{\Lambda}\nabla \cdot e^\psi\, v_i - \frac{1}{2}\int_\Omega f^{u_2} v_i \qquad 4.3$$

$$r_i(p) = \int_\Omega (\partial_1 u_1 + \partial_2 u_2)\, \varphi_i \qquad 4.4$$

$$r_i(\psi_{11}) = \int_\Omega \frac{\psi_{11} - \psi_{11}^n}{\Delta t} \chi_i + \int_\Omega (u_1\partial_1 \psi_{11} + u_2\partial_2 \psi_{11})\, \chi_i$$
$$- \int_\Omega (\omega_{11}\psi_{11} + \omega_{12}\psi_{21} - \psi_{11}\omega_{12} - \psi_{12}\omega_{12})\, \chi_i \qquad 4.5$$
$$- 2\int_\Omega B_{11}\, \chi_i - \int_\Omega \frac{1}{\Lambda}(\psi_{11} - 1)\, \chi_i - \int_\Omega f^{\psi_{11}} \chi_i$$

$$r_i(\psi_{12}) = \int_\Omega \frac{\psi_{12} - \psi_{12}^n}{\Delta t} \chi_i + \int_\Omega (u_1\partial_1 \psi_{12} + u_2\partial_2 \psi_{12})\, \chi_i$$
$$- \int_\Omega (\omega_{11}\psi_{12} + \omega_{12}\psi_{22} - \psi_{11}\omega_{12} - \psi_{12}\omega_{22})\, \chi_i \qquad 4.6$$
$$- 2\int_\Omega B_{12}\, \chi_i - \int_\Omega \frac{1}{\Lambda}(\psi_{12})\, \chi_i - \int_\Omega f^{\psi_{12}} \chi_i$$

4.2. THE JACOBIAN MATRIX

$$r_i(\psi_{22}) = \int_\Omega \frac{\psi_{22} - \psi_{22}^n}{\Delta t} \chi_i + \int_\Omega (u_1 \partial_1 \psi_{22} + u_2 \partial_2 \psi_{22}) \chi_i$$

$$- \int_\Omega (\omega_{21} \psi_{12} + \omega_{22} \psi_{22} - \psi_{21} \omega_{12} - \psi_{22} \omega_{22}) \chi_i \qquad 4.7$$

$$-2 \int_\Omega B_{22} \chi_i - \int_\Omega \frac{1}{\Lambda} (\psi_{22} - 1) \chi_i - \int_\Omega f^{\psi_{22}} \chi_i$$

The Jacobian matrix structure is,

$$\mathbf{J} = \frac{\partial \mathbf{r}_i}{\partial \mathbf{x}_j} = \begin{pmatrix} \frac{\partial r_i(u_1)}{\partial u_{1j}} & \frac{\partial r_i(u_1)}{\partial u_{2j}} & \frac{\partial r_i(u_1)}{\partial \psi_{11j}} & \frac{\partial r_i(u_1)}{\partial \psi_{12j}} & \frac{\partial r_i(u_1)}{\partial u_{22j}} & \frac{\partial r_i(u_1)}{\partial p_j} \\ \frac{\partial r_i(u_2)}{\partial u_{1j}} & \frac{\partial r_i(u_2)}{\partial u_{2j}} & \frac{\partial r_i(u_2)}{\partial \psi_{11j}} & \frac{\partial r_i(u_2)}{\partial \psi_{12j}} & \frac{\partial r_i(u_2)}{\partial u_{22j}} & \frac{\partial r_i(u_2)}{\partial p_j} \\ \frac{\partial r_i(\psi_{11})}{\partial u_{1j}} & \frac{\partial r_i(\psi_{11})}{\partial u_{2j}} & \frac{\partial r_i(\psi_{11})}{\partial \psi_{11j}} & \frac{\partial r_i(\psi_{11})}{\partial \psi_{12j}} & \frac{\partial r_i(\psi_{11})}{\partial u_{22j}} & \frac{\partial r_i(\psi_{11})}{\partial p_j} \\ \frac{\partial r_i(\psi_{12})}{\partial u_{1j}} & \frac{\partial r_i(\psi_{12})}{\partial u_{2j}} & \frac{\partial r_i(\psi_{12})}{\partial \psi_{11j}} & \frac{\partial r_i(\psi_{12})}{\partial \psi_{12j}} & \frac{\partial r_i(\psi_{12})}{\partial u_{22j}} & \frac{\partial r_i(\psi_{12})}{\partial p_j} \\ \frac{\partial r_i(\psi_{22})}{\partial u_{1j}} & \frac{\partial r_i(\psi_{22})}{\partial u_{2j}} & \frac{\partial r_i(\psi_{22})}{\partial \psi_{11j}} & \frac{\partial r_i(\psi_{22})}{\partial \psi_{12j}} & \frac{\partial r_i(\psi_{22})}{\partial u_{22j}} & \frac{\partial r_i(\psi_{22})}{\partial p_j} \\ \frac{\partial r_i(p)}{\partial u_{1j}} & \frac{\partial r_i(p)}{\partial u_{2j}} & \frac{\partial r_i(p)}{\partial \psi_{11j}} & \frac{\partial r_i(p)}{\partial \psi_{12j}} & \frac{\partial r_i(p)}{\partial u_{22j}} & \frac{\partial r_i(p)}{\partial p_j} \end{pmatrix} \qquad 4.8$$

In order to compute the Jacobian matrix components, we recall

$$u_1^h = \sum_{j=1}^{NN_{u_1}} u_{1j} \, v_j, \quad u_2^h = \sum_{j=1}^{NN_{u_2}} u_{2j} \, v_j, \quad p^h = \sum_{i=1}^{NN_p} p_j \, \varphi_j$$

$$\psi_{11}^h = \sum_{i=1}^{NN_{\psi_{11}}} \psi_{11j} \, \chi_j, \quad \psi_{12}^h = \sum_{i=1}^{NN_{\psi_{12}}} \psi_{12j} \, \chi_j, \quad \psi_{22}^h = \sum_{i=1}^{NN_{\psi_{22}}} \psi_{22j} \, \chi_j$$

4.9

with NN is the total node number in each variable. Thus,

$$\frac{\partial u_1^h}{\partial u_{1j}} = v_j, \quad \frac{\partial \nabla u_1^h}{\partial u_{1j}} = \nabla v_j, \quad \frac{\partial u_1^h}{\partial p_j} = \frac{\partial u_1^h}{\partial \psi_{11j}} = \frac{\partial u_1^h}{\partial \psi_{12j}} = \frac{\partial u_1^h}{\partial \psi_{22j}} = 0, \text{etc.} \qquad 4.10$$

If we have enough attention to finish computing all components by hand, this technique might results in highly accurate and optimized simulations, but unfortunately, even if the analytical expressions are available, it still needs a lot of efforts to construct the full Fréchet derivative discretization. For example in the case of Newtonian flow, the first two components are given by

$$\frac{\partial r_i(u_1)}{\partial u_{1j}} = \int_\Omega \frac{1}{\Delta t} \, v_j v_i + \frac{1}{2} \int_\Omega (v_j \partial_1 u_1 + u_1 \partial_1 v_j + u_2 \partial_2 v_j) \, v_i$$

$$+ \frac{1}{2} \int_\Omega \eta_s \nabla v_j \cdot \nabla v_i \qquad 4.11$$

$$\frac{\partial r_i(u_1)}{\partial u_{2j}} = \frac{1}{2} \int_\Omega (v_j \partial_2 u_1) \, v_i$$

As far as one can see, this is not a straightforward differentiation technique. Thus, one wants to avoid this technique for code flexibility reason in dealing with many different non-linear equations especially for viscoelastic models in the coming future.

4.2.2 Inexact Jacobian

Another well-known and widely used approach is to approximate the Jacobian, $\mathbf{J}(\mathbf{x}^n)$, with divided difference method. The advantage of this method is that it serves in a black box

CHAPTER 4. THE NUMERICAL SOLVER

manner so that it allows any nonlinear equations to be handled automatically without having to derive the corresponding analytical expressions, if existing at all. The method is described as follows,

$$\left[\frac{\partial r(\mathbf{x}^n)}{\partial \mathbf{x}}\right]_{ij} \approx \frac{r_i(\mathbf{x}^n + \epsilon_j \delta_j) - r_i(\mathbf{x}^n - \epsilon_j \delta_j)}{2\epsilon_j} \qquad 4.12$$

where δ_j is the basis unit vectors that determine the direction of the derivative and ϵ_j is a suitable step size. Other advantage of this method is that the structure of the Jacobian matrix is known a priori, which is typically the same as the nonlinear operator in equation (4.1). The function evaluation can then be done in the same way. The only minor thing is the dependency of its accuracy on the chosen step size, i.e. the step size must be small enough to avoid round-off effects arising from this difference method, equation (4.12), and at the same time should be large enough to avoid a truncation error. In the theory, the optimal step should be $\epsilon^{\frac{1}{3}} \approx 10^{-6}$ for double machine precision. Yet in practice the step size is chosen to be $\epsilon^{\frac{1}{2}} \approx 10^{-8}$.

Algorithm 2: Jacobian approximation

---
    forall the *Elements* **do**
1   Compute transformation from the reference element
2   Compute basis functions in the cubature points
       forall the *Cubature points* **do**
3   Compute the integrands
4   Do the difference
5   Add them to the local matrix
6   Update the global matrix
7   end

The algorithm above shows how to prepare the Jacobian matrix. In general, one prepares the basis/test functions for all elements and integration rules for all degrees of freedoms. Then, the computation in every element can be separated into degrees of freedoms that lay in the domains Ω (2d quadrature rule) and on the boundary Γ (1d quadrature rule).

Once a linear system is solved (later in the next section) within one Newton step (step 4 in Algorithm 1), then the next task is to find a damping parameter ω that should be able to ensure the convergence of the Newton. As already mentioned, this study implements a line search technique which is based on the backtracking idea, see [24, 59]. Here, a full step ($\omega = -1$) is checked whether $\mathbf{r}(\omega) = \mathbf{r}(\mathbf{X}^n + \omega \delta \mathbf{X})$ is minimized or not. If the starting solution is close then this is mostly satisfying. If it is not, then at least we can save this information and use it for searching a new damping factor ω by backtracking along the Newton direction $\delta \mathbf{X}$ until $\mathbf{r}(\omega) = \mathbf{r}(\mathbf{X}^n + \omega \delta \mathbf{X})$ is minimized. To achieve this goal, a quadratic interpolation is used to obtain the corresponding ω since we have 3 informations already: $\mathbf{r}(\omega = -1)$ from the full step test, $\mathbf{r}(0)$ from step 2 of algorithm 1, and $\mathbf{r}'(0)$ from step 3 algorithm 1. This can be visualized by Fig. 4.1 as 1D problem. Then, the new damping factor reads

$$\omega_1 = \frac{-\mathbf{r}'(0)}{2(\mathbf{r}(1) - \mathbf{r}(0) - \mathbf{r}'(0))}. \qquad 4.13$$

By replacing $\mathbf{r}(\omega = -1)$ with the last computed information $\mathbf{r}(\omega_1)$, the next damping parameter can be iteratively computed in the same way until $\mathbf{r}(\omega_n)$ meets the criterion.

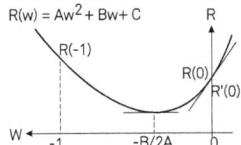

Figure 4.1: The quadratic interpolation.

4.3 Inner multigrid solver

Consider that one has the Jacobian matrix at hand. Now, the 'big' task, which means the most time consuming part of the processing, is to solve the linear subproblems (step 4 in Algorithm 1). Given a sparse matrix $\mathbf{J}$, which contains mostly zeros element, one needs to solve for $\delta \mathbf{X}$,

$$\mathbf{J}(\mathbf{x})\delta \mathbf{x} = \mathbf{r}(\mathbf{x}). \qquad 4.14$$

If high computer memory is provided, then a direct solver for sparse systems might be in favor. This can be Gauss elimination, LU decomposition, or cyclic reduction technique. The already available routine is UMFPACK (Unsymmetric MultiFrontal Package) which can solve a sparse matrix problem with reduced fill-in element, see [23]. But if computer memory is limited, then iterative methods would be the best choice. This can be ILU decomposition, alternating direction implicit (ADI) splitting method, conjugate gradient method, biconjugate gradients (a modified conjugate gradient method) and variants (BICG, BICGSTAB, GMRES) or multigrid methods. Each method has its own limitations, i.e. conjugate gradient method needs a positive definite matrix. Among these methods, multigrid offers a rather different approach and substantial benefits such as it is potentially most efficient in terms of computational cost, see [75].

It seems that nowadays 'Multigrid' has been a keyword for highly efficient iterative solvers in handling the linear subproblem described in the algorithm above (step 3). It can be true if multigrid is 'optimally' implemented. Multigrid components consist of smoother, restriction, prolongation and a direct coarse grid solver (UMFPACK). The way multigrid solves a discrete system is facilitated by the construction of lower levels of discrete systems, which is known as standard geometric multigrid. Inside the multigrid solver, few smoothing steps are needed before and after both restriction and prolongation step. The applied smoother is of Vanka-type,

$$\begin{bmatrix} \mathbf{u}^{l+1} \\ \psi^{l+1} \\ p^{l+1} \end{bmatrix} = \begin{bmatrix} \mathbf{u}^{l} \\ \psi^{l} \\ p^{l} \end{bmatrix} + \omega^{l} \sum_{T\in\mathcal{T}_h} [\mathbf{J}]^{-1}_{|T} \begin{bmatrix} Res_{\mathbf{u}} \\ Res_{\psi} \\ Res_{p} \end{bmatrix}_{|T}, \qquad 4.15$$

which shows an effective coupled way of solving with local pressure Schur complement approach. The second term of the right hand side of equation (4.15) acts locally in each element T on all levels. Here, the 'summation' over each element represents an assembling technique and the inverse of the local systems on each element T (in 2D case of size 48×48) is computed by a direct solver (LAPACK). After few number of smoothing steps, an approximation of the solution is obtained $\mathbf{u}^h$. The unknown error of this approximation ($\mathbf{e}^h = \mathbf{u} - \mathbf{u}^h$) has been smoothed. So, the restriction sets in to approximate error on the coarser level $\mathbf{e}^{2h}$, which will then be smoothed again. This action continues until we obtain the smoothed error on the coarsest grid, $\mathbf{e}^c$. A 'V-cycle' will visit the coarsest grid once and a 'F-cycle' may do it a few times more. At the coarsest grid, the linear system of the residual equation $\mathbf{J}(\mathbf{x}^n)\mathbf{e}^c = \mathbf{r}(\mathbf{x}^n)$ is

CHAPTER 4. THE NUMERICAL SOLVER

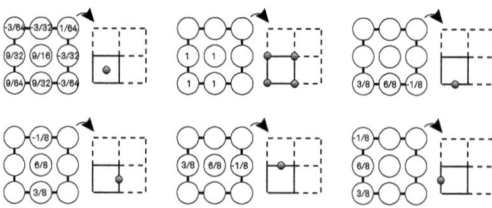

Figure 4.2: Prolongation in Q_2 with biquadratic interpolation.

solved by a direct sparse solver (UMFPACK). In order to be efficient, the coarsest grid should be coarse enough so that requirement of UMFPACK memory is as low as possible. Then, we apply prolongation which is followed by post-smoothing to give a better error approximation back to the finest level. These steps continue until a 'V or F-cycle' of multigrid iterations is finished. In the case of conforming finite element, the lower level space is a subspace of the finer level one. Thus, natural injection can be applied for the restriction operator. The prolongation operator, on the other hand, needs to be constructed by biquadratic interpolation, see Fig. 4.2. In general, one multigrid cycle is described by the algorithm 3 where the

---
Algorithm 3: One cycle multigrid

Input: linear tolerance
1. set min/max of level L
2. Call MG(L)/smooth $\mathbf{e}^L$
3. Restrict $\mathbf{A}, \mathbf{r}$
   if *(L-1=min)* then
4. | Solve $\mathbf{Ae} = \mathbf{r}$
5. | Prolong $\mathbf{e}$
   else
6. | Call MG($L-1$)/smooth $\mathbf{e}^{L-1}$
7. end

---

recursive call MG is under taken if the coarse multigrid level is not yet reached.

4.4 Solver behavior

It is well-known that Newton and multigrid solvers are sensitive with respect to the choice of parameter settings, i.e. in the case of multigrid: number of smoothing steps, prolongation damping, stopping criterion, etc. and in the case of Newton: line search and Jacobian parameter settings. These can be different from one to another problem. Yet, in the monolithic treatment of the non-Newtonian fluids given in the nonlinear viscosity functions, the proposed Newton-multigrid approach behaves very well as described in the following subsection.

4.4.1 Nonlinear viscosity in flow around cylinder configuration

We consider steady fluid flow problems of the generalized Newtonian flow that satisfy

$$\begin{cases} (\mathbf{u} \cdot \nabla)\mathbf{u} = -\nabla p + \nabla \cdot \mathbf{T} \\ \nabla \cdot \mathbf{u} = 0 \end{cases} \qquad 4.16$$

4.4. SOLVER BEHAVIOR

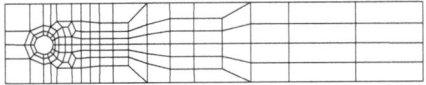

Lev.	NEL	NVT	NMP
L1	520	572	1092
L2	2080	2184	4264
L3	8320	8528	16848
L4	33280	33696	66976
L5	133120	133952	267072

Figure 4.3: Flow around a cylinder coarse mesh. NEL, NVT and NMP are number of element, vertices and mid points.

with the constitutive law $\mathbf{T} = 2\eta(\dot{\gamma}, p)\mathbf{D}$ and $\dot{\gamma}$ is the magnitude of the deformation tensor. The nonlinear viscosity function follows the generalized Cross model with additional pressure dependant and bounds the viscosity given by

$$\eta(\dot{\gamma}, p) = \eta_{\min} + \frac{\eta_{\max} - \eta_{\min}}{(1 + \lambda\sqrt{\mathrm{tr}(\mathbf{D}^2)})^r} \exp(\alpha_c p). \qquad 4.17$$

The computational domain can be seen in Fig. 4.3 with the corresponding mesh information for 5 refinement levels. Inflow is set to a parabolic profile of $u_x(y) = 0.3 \times 4.0/0.1681\, y\, (0.41 - y)$ and lower bound viscosity is always set to $\eta_{\min} = 10^{-3}$. Functional value of drag forces, F_{drag}, are computed around the cylinder surface s [73],

$$F_{\mathrm{drag}} = \int_s (-p\mathbf{I} + \mathbf{T}) \cdot \mathbf{n}\, ds, \qquad 4.18$$

for different mesh levels. The parameters of r, α_c and $\eta_{\max}$ are set in such a way that steady state solutions can still be obtained. Undesired parameter settings that lead to nonstationary solutions behavior are avoided. In the following Table 4.1, the monolithic approach shows stable Newton-multigrid convergence rates in terms of number of nonlinear (NN) and linear multigrid sweep (LL) iterations. Here, the linear multigrid gains 2 digits linear tolerance with 4 smoothing steps. The outer Newton satisfies an error of 10^{-8} and the divided differences Jacobian step length is set to the double machine precision.   Mesh converged solutions

Table 4.1: **Cross-like model:** Newton-Multigrid behavior in term NL/LL [F_{drag}]. Initial solutions are zero vector for L1, while initial solutions for other levels use the solution of one level below.

Lev.	$r = 1, \alpha_c = 0$		
	$\eta_{max} = 10^{-2}$	$\eta_{max} = 10^{-1}$	$\eta_{max} = 1$
L1	9/1 [1.2565e-2]	11/2 [2.972e-2]	30/2 [0.20942]
L2	3/1 [1.2626e-2]	4/2 [2.9832e-2]	26/3 [0.20890]
L3	3/2 [1.2647e-2]	4/2 [2.9874e-2]	24/3 [0.20876]
L4	3/2 [1.2652e-2]	4/3 [2.9885e-2]	16/3 [0.20875]
L5	2/2 [1.2654e-2]	3/2 [2.9888e-2]	14/3 [0.20876]
	$r = 1, \alpha_c = 1$		
L1	11/1 [1.2574e-02]	12/2 [3.0215e-02]	15/3 [2.6196e-01]
L2	3/1 [1.2635e-02]	4/3 [3.0327e-02]	31/4 [2.5983e-01]
L3	3/2 [1.2656e-02]	4/3 [3.0370e-02]	31/7 [2.5934e-01]
L4	3/2 [1.2662e-02]	4/3 [3.0382e-02]	13/5 [2.5927e-01]
L5	2/2 [1.2663e-02]	3/3 [3.0385e-02]	17/3 [2.5928e-01]

CHAPTER 4. THE NUMERICAL SOLVER

are obtained for 5 refinement levels and the Newton-multigrid solver behaves very well with respect to the chosen parameter settings.

In the following, Newton-multigrid solver are tested on viscoelastic problems. In order to ease multigrid parameter settings, especially on hard problems such as high Re or We numbers, a stabilization technique (it is described in previous chapter) is used. At the same time, it can also ease the already described mixed formulation problem. This is a well-known issue due to the native hyperbolic type of the given viscoelastic formulation. Therefore, a stabilization technique is unavoidable. We present in the following some numerical experiments with artificial diffusion, which can help the linear solver task, given in the flow around cylinder viscoelastic benchmark problem, see Fig. 4.4.

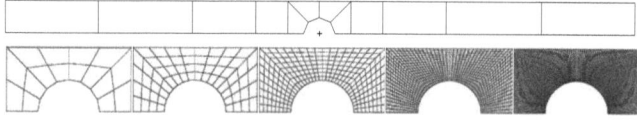

Figure 4.4: Top: The coarse mesh. Bottom: A sequence of 5 level refinements.

4.4.2 Flow around a cylinder with artificial diffusion

Here, it is shown that a simple artificial diffusion may help multigrid to converge properly. This additional stabilization term acts in the form of $\gamma h \Delta \tau$ to the original problem, which is solved by direct steady calculation,

$$\begin{cases} \nabla p - 2\eta_s \Delta \mathbf{u} - \dfrac{\eta_p}{\Lambda} \nabla \cdot \boldsymbol{\tau} = 0 \\ \nabla \cdot \mathbf{u} = 0 \\ (\mathbf{u} \cdot \nabla)\boldsymbol{\tau} - \nabla \mathbf{u} \cdot \boldsymbol{\tau} - \boldsymbol{\tau} \cdot \nabla \mathbf{u}^T + \dfrac{1}{\Lambda}(\boldsymbol{\tau} - \mathbf{I}) + \gamma h \Delta \boldsymbol{\tau} = 0 \end{cases} \quad 4.19$$

The parameter γh is elementwise localized, which has a tendency to decrease by increasing refinement. This is slightly similar to the work of Sureshkumar and Beris [66]. The only difference is that they set the artificial diffusion parameter as a global constant without dependency of h. A fully developed velocity profile is set at the inflow, $u_x = 1.5(1 - \frac{y^2}{4})$, and the same velocity profile is obtained at the outflow by setting Neumann boundary data. The full domain is geometrical symmetry. Thus, by assuming steady flow state (low We number) we can take half of the domain in the numerical computation, as shown in Fig. 4.4.

In the following numerical tests, two different We = 0.1 and We = 0.7 numbers are tested in the above mentioned configuration. Although for the latter We number it is still debatable whether the solution reaches steady data or not, this work claims it does (later discussed in sub chapter 5.5). Table 4.2 shows that multigrid works well with artificial diffusion. The drag coefficient may not be the same as the one in the benchmark (see [21]) or later in chapter 5, but apparently it can be improved by reducing the coefficient γ. This optimum coefficient ($\gamma = 0.25$, see the third sub table of Tab. 4.2) gives a desired convergence rate while at the same time gives an acceptable accuracy in term of drag coefficient.

In order to maintain the originality of the problem and thus to get closer numbers of drag coefficient, one may set different coefficients γ for the Jacobian matrix (γ_j) and the residual (γ_r) in such a way $\gamma_r \leq \gamma_j$. This means that we set different parameter to the discretized matrix of left hand side (Jacobi) than to the discretized matrix of the right hand side (residual). By setting $\gamma_r = 0$, one recovers the original equation which is without extra

4.4. SOLVER BEHAVIOR

Table 4.2: **Oldroyd-B, Conformation:** Newton-Multigrid behavior in term NL/LL [Drag]

	$\gamma=1.0$		$\gamma=0.5$	
Lev	We = 0.1	We = 0.7	We = 0.1	We = 0.7
L1	5/1 [106.9872]	6/1 [87.0798]	5/1 [113.7894]	8/1 [92.7863]
L2	3/1 [115.6485]	4/2 [94.3514]	3/1 [120.871]	4/2 [100.2350]
L3	3/2 [121.6353]	4/3 [100.827]	3/1 [125.197]	4/2 [106.1147]
L4	3/2 [125.422]	4/4 [106.2924]	3/2 [127.6223]	4/3 [110.4968]
L5	3/2 [127.6816]	4/6 [110.5438]	2/2 [128.9401]	4/4 [113.5349]
	$\gamma=0.25$		$\gamma_{j/r} = 0.25/0.15$	
Lev	We = 0.1	We = 0.7	We = 0.1	We = 0.7
L1	5/1 [118.9325]	9/2 [98.6612]	11/1 [121.6486]	23/1 [102.6969]
L2	3/1 [124.4178]	4/2 [105.5058]	7/1 [126.1438]	12/1 [108.7349]
L3	3/1 [127.3941]	4/2 [110.3037]	6/1 [128.3901]	14/1 [112.6077]
L4	2/2 [128.8801]	3/3 [113.4846]	6/1 [129.4218]	14/2 [115.0053]
L5	2/2 [129.6212]	3/4 [115.4835]	6/2 [129.9049]	14/2 [116.4005]
	$\gamma_{j/r} = 0.25/0.1$		$\gamma_{j/r} = 0.25/0.05$	
Lev	We = 0.1	We = 0.7	We = 0.1	We = 0.7
L1	16/1 [123.2419]	37/1 [105.6800]	26/1 [125.0641]	93/1 [111.2205]
L2	12/1 [127.1039]	29/1 [110.8638]	23/1 [128.1489]	61/1 [113.6914]
L3	7/1 [128.9207]	21/1 [114.0073]	25/1 [129.4767]	57/1 [115.6824]
L4	8/1 [129.7022]	23/1 [115.8786]	10/1 [129.9895]	41/1 [116.8352]
L5	8/1 [130.0492]	23/2 [116.8846]	12/1 [130.1952]	48/1 [117.3436]

diffusion term,

$$(\mathbf{u}\cdot\nabla)\boldsymbol{\tau} - \nabla\mathbf{u}\cdot\boldsymbol{\tau} - \boldsymbol{\tau}\cdot\nabla\mathbf{u}^T + \frac{1}{\Lambda}(\boldsymbol{\tau}-\mathbf{I}) + \gamma h \Delta\boldsymbol{\tau} = 0. \qquad 4.20$$

Nevertheless, just by decreasing γ_r a satisfied drag coefficient can be reached with approximately the same multigrid rate (see the last sub table of Tab. 4.2). The price is that the nonlinear iteration increases moderately due to the disturbed Jacobian **J**. Yet, the multigrid behaves optimal with respect to mesh refinement.

In the following example we show that the proposed monolithic approach is independent of the given formulation and also performs very well with the LCR, which has an exponential term

$$\begin{cases} \nabla p - 2\eta_s \Delta \mathbf{u} - \dfrac{\eta_p}{\Lambda} \nabla \cdot e^{\boldsymbol{\psi}} = 0 \\ \nabla \cdot \mathbf{u} = 0 \\ (\mathbf{u}\cdot\nabla)\boldsymbol{\psi} - (\boldsymbol{\Omega}\boldsymbol{\psi} - \boldsymbol{\psi}\boldsymbol{\Omega}) - 2\mathbf{B} + \gamma h \Delta \boldsymbol{\psi} = \dfrac{1}{\Lambda} f(\boldsymbol{\psi}) \end{cases} \qquad 4.21$$

Tab. 4.3 shows us that the linear solver behaves the same as the one with the conformation stress tensor formulation in Tab. 4.2. The linear convergence rate is independent of the chosen We number and the drag results are similar. Both Newton-multigrid solvers are stable with respect to mesh refinement. Nonlinearity of the problem is clearly shown by the increasing number of nonlinear iteration at higher We number. Although the drag coefficient may not be the same as the one with conformation tensor due to different stabilization terms ($\gamma h \Delta \boldsymbol{\tau}$ and $\gamma h \Delta \boldsymbol{\psi}$), the solver behavior remains stable for both cases.

CHAPTER 4. THE NUMERICAL SOLVER

Table 4.3: **Oldroyd-B, LCR:** Newton-Multigrid behavior in term NL/LL [Drag]

	$\gamma=1.0$		$\gamma=0.5$	
Lev	We = 0.1	We = 0.7	We = 0.1	We = 0.7
L1	6/1 [105.5171]	6/1 [83.2051]	6/1 [112.1263]	7/1 [86.7057]
L2	3/1 [113.7893]	3/2 [88.0028]	3/1 [119.0820]	3/1 [92.1309]
L3	3/1 [119.8001]	3/2 [92.6686]	3/1 [123.6846]	3/2 [97.2610]
L4	3/2 [123.9020]	4/3 [97.4222]	3/2 [126.5245]	3/3 [102.1937]
L5	3/2 [126.5826]	3/4 [102.2374]	3/2 [128.2260]	3/3 [106.6273]
	$\gamma=0.25$		$\gamma_{j/r}=0.25/0.15$	
Lev	We = 0.1	We = 0.7	We = 0.1	We = 0.7
L1	6/1 [117.4035]	7/2 [90.8816]	9/1 [120.3768]	13/1 [94.3513]
L2	3/1 [122.9492]	3/2 [96.7072]	8/1 [124.9818]	9/1 [100.2434]
L3	3/1 [126.3008]	4/2 [102.0235]	6/1 [127.5860]	8/2 [105.3236]
L4	3/2 [128.1667]	3/2 [106.5813]	5/1 [128.9289]	7/2 [109.2731]
L5	3/2 [129.1929]	3/3 [110.1557]	4/2 [129.6219]	7/3 [112.1590]
	$\gamma_{j/r}=0.25/0.1$		$\gamma_{j/r}=0.25/0.05$	
Lev	We = 0.1	We = 0.7	We = 0.1	We = 0.7
L1	13/1 [122.2221]	19/1 [97.3749]	22/1 [124.4786]	36/1 [103.2290]
L2	14/1 [126.1832]	15/1 [103.0146]	20/1 [127.5802]	34/1 [107.3007]
L3	6/1 [128.3114]	13/2 [107.6439]	18/1 [129.1174]	23/1 [110.8732]
L4	7/1 [129.3435]	12/2 [111.0432]	10/1 [129.7888]	23/2 [113.3784]
L5	6/1 [129.8491]	11/3 [113.4227]	10/1 [130.0873]	20/2 [115.0071]

4.4.3 Viscoelastic flow in a cavity with EO-FEM stabilization

The flow around cylinder configuration is probably a difficult example. Let us take a moderate example with the cavity configuration as seen in Fig. 4.5. Here, LCR is used for both Oldroyd-B and Giesekus models and EO-FEM is applied for the stabilization of numerics. The set up of the problem follows the one in [58] where a velocity data of value $u_x = 1$ is set to the top boundary data and no-slip condition is set for the others. A direct steady simulation up to We = 1.0 is undergone to see the solver behavior. The simulation becomes unsteady above We = 1.0 in the cavity configuration, which is later described in chapter 6.

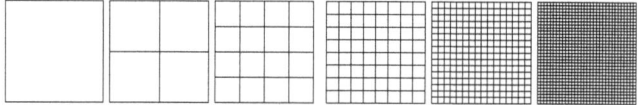

Figure 4.5: Sequence of mesh refinement for cavity

Table 4.4 shows the behavior of Newton-multigrid in a viscoelastic simulation inside a cavity with 2-grid algorithm. Here, the starting solution is We = 0.1 at L3. A constant damping factor of 0.8 and number of smoothing steps of 4 are being used. L4 solutions are computed by initial solution of L3, and L5 is solved by following the same way. It is shown that for both viscoelastic models the solver remains stable, with respect to increasing We number, with maximum of 4 linear multigrid sweeps. Although there is still tendency that the number of linear multigrid sweep increases by refinement. This must be further studied. The cause might come from the setup of configuration (cavity) with everywhere Dirichlet data and high order conforming finite element Q_2, or might be the chosen multigrid parameters

4.4. SOLVER BEHAVIOR

Table 4.4: **Oldroyd-B (Old) and Giesekus (Gie) model:** Newton-Multigrid behavior in term NL/LL with 2-grid algorithm.

We	L3	L4	L5	L3	L4	L5
	Old			Gie		
0.2	6/1	4/2	4/3	5/1	4/2	4/3
0.3	5/1	5/2	4/3	4/1	4/2	4/3
0.4	5/1	5/2	5/4	4/1	4/2	4/3
0.5	5/2	5/3	6/4	4/1	4/2	4/3
0.6	5/2	5/3	5/4	3/2	4/2	4/3
0.7	5/2	5/3	5/4	4/1	4/2	4/3
0.8	5/2	5/3	6/3	5/1	4/2	4/3
0.9	5/2	5/3	6/3	4/1	4/2	4/3
1.0	5/2	6/3	6/4	4/1	4/2	4/4

setting which are not suitable for the corresponding cavity configuration. Many things can trigger this phenomena. In the following, multigrid parameter settings are played around. Tab.4.5 shows that by increasing the number of smoothing steps, the load of linear multigrid sweeps can be reduced. Here, there is tendency that increasing smoothing steps may let 2-grid algorithm independent of refinement. Multigrid parameters that may be further studied are linear tolerance, damping of smoother and prolongation.

Table 4.5: **Oldroyd-B (Old) and Giesekus (Gie) model:** Newton-Multigrid behavior in term NL/LL with 2-grid algorithm and varied smoothing steps.

We	L3	L4	L5	L3	L4	L5
	Old			Gie		
ss=3						
0.2	6/1	5/2	4/3	4/1	3/2	4/3
0.6	5/2	5/3	6/4	4/2	4/3	4/4
1.0	5/2	5/4	9/3	4/2	4/3	4/5
ss=7						
0.2	5/1	5/1	5/2	6/1	4/1	3/2
0.6	5/1	6/2	5/3	4/1	4/2	4/3
1.0	5/1	5/2	5/3	3/1	4/2	4/2
ss=17						
0.2	5/1	5/1	4/1	5/1	3/1	4/1
0.6	5/1	5/1	5/2	4/1	4/1	4/2
1.0	5/1	5/1	6/2	3/1	4/1	4/2

Now, we learn that solving numerical problem with multigrid is to find the best multigrid parameter for each individual problem, especially when the problem is highly nonlinear.

5
Numerical Validation

This chapter is dedicated to the validation of the code, which is a very crucial part of the thesis. Here, several well-known benchmarks are performed step by step towards the final target of solving viscoelastic flow problems. The validation shows also the quality of the high order finite element space (Q_2/P_1) within the monolithic coupled approach with a strong Newton solver. The benchmarks are without doubt very interesting for the research and important as well for the industrial purposes. The first benchmark test is the flow around cylinder, where drag and lift values are computed. The second is cavity flow, where kinetic energy is taken into consideration. The above benchmarks solve the Navier-Stokes equations. Standing vortex problem is then presented to show the potential of EO-FEM when solving Navier-Stokes equations without a diffusive term. The next benchmark is about nonisothermal convection flow in a stretched 8:1 cavity configuration, which is well known as the MIT benchmark 2001. The Navier-Stokes equations plus an energy equation are solved within this benchmark. The Nusselt number is computed and compared between different authors and methods. This benchmark is very advantageous considering many contributors that participated in. The last benchmark is for the viscoelastic benchmark which presents the drag and stress plot in a planar flow around a cylinder for both Oldroyd-B and Giesekus type of fluid.

5.1 Flow around cylinder benchmark

This benchmark was started within the DFG high-priority research program "flow simulation with high performance computers" by Schäfer and Turek [73]. It describes the flow around a cylinder situation, which in 2D looks like as in Fig. 5.1. The cylinder is not located on the symmetry line but rather slightly shifted from it. This is to trigger unsteady flow behavior and the build up of 'Von Karman vortex shedding' above a certain Re number. Inflow is set to

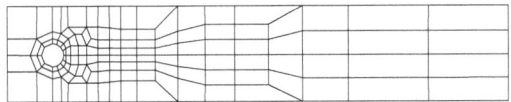

Figure 5.1: Flow around a cylinder coarse mesh L0, cells=130.

a parabolic profile $u_x(y) = 0.3 \times 4.0/0.1681 \, y \, (0.41 - y)$ with kinematic viscosity $\eta = 1e-3$. The mean velocity across the inflow is $u_c = 0.2$, and the characteristic length of the cylinder is $l_c = 0.1$. This setting corresponds to Re = 20. Functional values of drag C_{drag} and lift C_{lift}

CHAPTER 5. NUMERICAL VALIDATION

are computed around the cylinder surface s [73],

$$C_{\text{drag/lift}} = C \int_s \boldsymbol{\sigma} \cdot \mathbf{n} \, ds \qquad 5.1$$

with a constant $C = 2.0 F_{d/l}/(\rho u_c^2 l_c)$. In this benchmark, $F_{d/l}$ are set to 1 which ends up with the same constant C for both drag and lift. To calculate equation (5.1), there are 2 possibilities of integration namely the usual line integration and the volume integration [43]. The latter approach takes benefits from the integration over the whole domain in the finite element discretization,

$$C_{\text{drag/lift}} = -C(\langle \eta \nabla \mathbf{u}, \nabla v \rangle - \langle p, \nabla \cdot v \rangle) \qquad 5.2$$

with a test function $v \in \mathbf{V}_h$ as described in chapter 3.

Table 5.1: **Drag/Lift** Re = 20: NL denotes the number of nonlinear iteration. LL denotes the average number of multigrid iteration while LI and VO denote the line and volume integration. Initial solution is zero vector. Lower/upper bound of reference values [73] are $C_{\text{drag}} = 5.5700/5.5900$, $C_{\text{lift}} = 0.0104/0.0110$

Lev.	Drag/Lift (LI)	Drag/Lift (VO)	NL/LL	Ref.[67]
L1	5.5550/ 0.009498	5.5424/ 0.00945	9/2	(L4) 5.5707/0.01046
L2	5.5722/ 0.010601	5.5672/ 0.01047	9/2	(L5) 5.5747/0.01055
L3	5.5776/ 0.010616	5.5761/ 0.01057	9/1	(L6) 5.5771/0.01058
L4	5.5790/ 0.010618	5.5786/ 0.01060	8/1	(L7) 5.5783/0.01060

In this configuration, where the solution is quite smooth (Re = 20), a fully converged solution is easier obtained for the higher order finite element space (Q_2/P_1) than with the corresponding lower order finite element space $(\tilde{Q}_1/Q_0)$ with pressure separation techniques [67] that needs more refinements. This means that the code does not necessarily apply EO-FEM for such a test case. We will come into this in the next sections. Another interesting comparison in this benchmark is non steady flow at (Re = 100), but which will not be covered in this study.

5.2 Driven cavity benchmark

This benchmark has a very simple configuration (a square) to test new solution methods. The standard case is fluid that is given shear flow on the top of the square which represent problems that are frequently encountered in industrial processes. A constant velocity $u_x = 1$ is set to the upper wall and a no-slip boundary condition is set to all other walls. Constant viscosities of $\eta = 10^{-3}, \eta = 2 \times 10^{-4}$ and $\eta = 10^{-4}$ are set which correspond to Re = 1000, Re = 5000 and Re = 10000. The kinetic energy is computed on every mesh levels with

$$E_{\text{kin}} = \frac{1}{2} \|\mathbf{u}_h\|_{0,\Omega}^2 \qquad 5.3$$

and compared with the low order finite element space with pressure separation technique [67] for Re = 1000 and Re = 5000 in Tab. 5.2. The comparison shows that high order finite element needs two refinements less than the low order counter part for approximately the same accuracy of kinetic energy. Higher inertia flow, i.e. higher Re number needs more refinement to obtain mesh independent solution.

5.2. DRIVEN CAVITY BENCHMARK

Table 5.2: **Kinetic energy:** Comparison with reference results for Re = 1000 and Re = 5000.

cells	Re = 1000 kinetic	Ref.[67]	Re = 5000 kinetic	Ref.[67]	Re = 10000 kinetic
256	5.2454e-02		1.0860e-01		1.3790e-01
1024	4.5418e-02	4.8095e-2	6.1149e-02	6.1785e-2	8.1822e-02
4096	4.4590e-02	4.5828e-2	4.9571e-02	5.5969e-2	5.3464e-02
16384	4.4524e-02	4.4843e-2	4.7691e-02	5.0196e-2	4.8726e-02
65536	4.4519e-02	4.4592e-2	4.7465e-02	4.8020e-2	4.7845e-02
262144	4.4518e-02	4.4535e-2	4.7430e-02	4.7541e-2	4.7813e-02

This comparison shows the potential of high order finite element space, with respect to mesh refinement, to obtain mesh converged solutions. Fig. 5.2 shows us from another point a view the obtained solutions. Here, the cutline of velocity magnitude is presented on the center line of the domain where high gradient of velocity occurs. By increasing of refinements the velocity profile along this line converged also for Re = 10000.

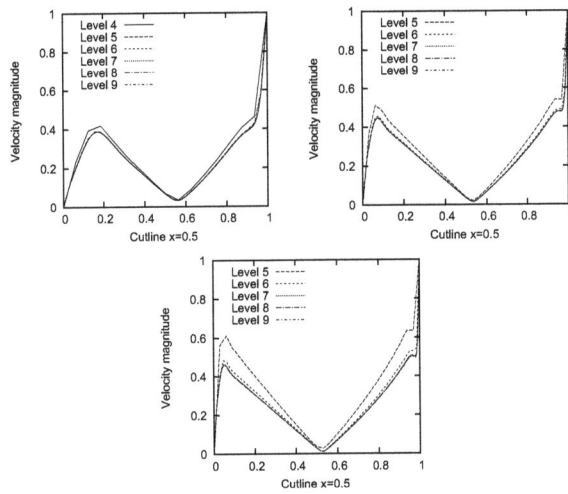

Figure 5.2: Velocity magnitude at center line. From left: Re = 1000, Re = 5000, Re = 10000

It is interesting to see from Fig. 5.3 that there appears a new vortex at the bottom right corner as inertia increases. A phenomena which is also experimentally known but numerically difficult in direct steady state calculation. Another approach to get a fully developed solution is by using non steady calculation. This is very often time consuming and computationally expensive.

The results of the first two benchmarks (flow around cylinder and driven cavity) in this study are in a good agreement with the reference values [67] and the code is validated for solving incompressible Navier-Stokes equation. Mesh converged solutions are obtained for

CHAPTER 5. NUMERICAL VALIDATION

Figure 5.3: Driven cavity flow. From left: Re = 1000, Re = 5000, Re = 10000

both benchmark problems. In the following section, we describe a more difficult problem and show the potential of EO-FEM as stabilization technique.

5.3 Standing vortex

The next validation is the well-known standing vortex problem [35], which solves an "infinite" Reynolds number inside a cavity. The problems are formulated as follows

$$\begin{cases} \rho \dfrac{\partial \mathbf{u}}{\partial t} + \rho (\mathbf{u} \cdot \nabla) \mathbf{u} = -\nabla p \\ \nabla \cdot \mathbf{u} = 0 \end{cases} \qquad 5.4$$

with an initial solution which looks like (in polar coordinate)

$$u_r = 0, \quad u_\theta = \begin{cases} 5r, & r < 0.2, \\ 2 - 5r, & 0.2 \le r \le 0.4, \\ 0, & r > 0.4, \end{cases} \qquad 5.5$$

where $r = \sqrt{(x-0.5)^2 + (y-0.5)^2}$ denotes the distance from the center. The initial solution represents also the steady-state solution of the above problem. By solving it with the time integration, the initial solution will evolve and change in shape. It is shown in [56] with lower order finite element spaces that without EO-FEM, the initial solution oscillates after time $t = 3$. As further reported in [56], the (first order) upwinding and also the streamline-diffusion method introduce too much artificial diffusion. In this study, where EO-FEM is implemented for high order finite element spaces, the same phenomena can be seen but at

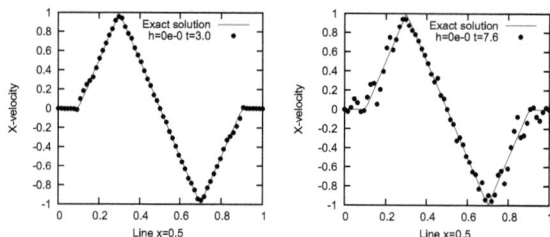

Figure 5.4: Standing vortex without EO-FEM stabilization

later time $t = 9$, see Fig. 5.4. The solution without EO-FEM at time $t = 3$ starts to deviate from the initial solution, which crash at time $t = 7.6$. The element pair suffers as well from instability for having no diffusion at all. By adding EO-FEM stabilization technique, the solutions remain the same even up to time $t = 9$, see Fig. 5.5 which shows two different EO-FEM parameters. The enormous potential of EO-FEM is then without doubt a clear message

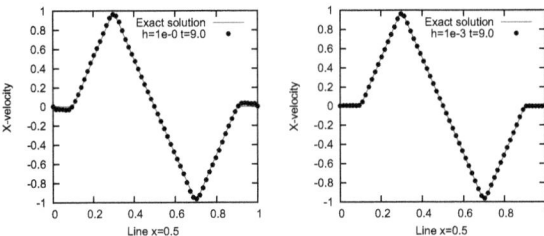

Figure 5.5: Standing vortex with EO-FEM stabilization

to any other convection-dominant flow such as in the case of viscoelastic flow problems.

5.4 MIT Benchmark 2001

The MIT Benchmark 2001 [16] describes a heat driven cavity flow in a 8:1 rectangular domain at near-critical Rayleigh number. Why near-critical? Because the onset of thermal convection will occur at the critical number, beyond that a non-periodical up to turbulent flow is resulting.

The geometry of the problem is very simple (see Fig. 5.6) but leads nevertheless to complex

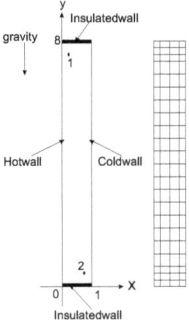

Figure 5.6: Geometry and coarse mesh

multi scale phenomena. The velocity vector at the upper and bottom wall is zero which describes a non-slip condition. The left wall is heated while the right wall is cooled by a prescribed non-dimensional temperature of -0.5 and 0.5. Gravity is applied downwards. The top and bottom of the walls are insulated, which means that homogeneous Neumann boundary conditions for the temperature are set and hence no heat is going outside of the wall. The initial condition is the zero vector for all variables. Physically relevant variables which are to

CHAPTER 5. NUMERICAL VALIDATION

Table 5.3: Contributor's and our testing meshes

Author	Turek	Davis	Gresho	Le Quéré
Mesh	128 x 704	83 x 403	105 x 481	48 x 180
Mesh	Elements	Nodes	Edges	Dof
R2a0	1408	1513	2920	21747
R2a1	1936	2043	3978	29679
R2a5	17776	17891	35666	267327
R3a0	5632	5841	11472	85731
R3a1	6688	6899	13586	101583
R3a4	21472	21689	43160	323379
R4	22528	22945	45472	340419
R4a1	24640	25059	49698	372111
R4a3	37312	37735	75046	562215

be computed are the velocity and temperature at point 1, and the Nusselt number along both sides of the wall. The time step is chosen so that there are enough data points in one oscillation of the resulting variables to graphically post process all quantities and so that smaller time steps do not significantly improve the solutions with respect to quantitative measurements. After comparison with the results from Davis [22], Gresho [36], Turek [62] and Le Quéré [64] we choose approximately 34 time steps in one oscillation which corresponds to $\Delta T = 0.1$ as time step size. Several meshes have been used to perform the spatial discretization (see Tab.5.3). The coarse mesh has approximately 1:5 x-to-y ratio of grid points and decreases gradually towards the walls (see Fig. 5.7).

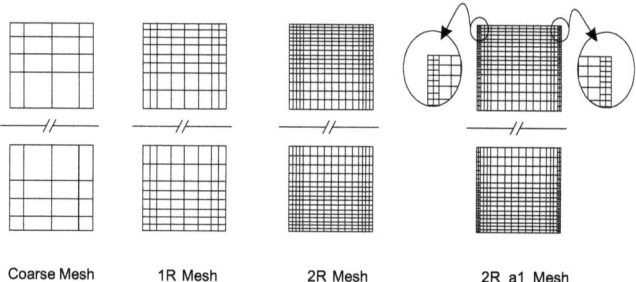

Figure 5.7: Several hierarchies and types of meshes

This figure also describes how the local refinement is generated for some exemplary meshes. We set the local refinement to be at both sides of the wall. This judgment is subject to the Nusselt number which is of interest for the engineer,

$$\text{Nu}(t) = \frac{1}{H} \int_0^H |\frac{\partial \Theta}{\partial x}|_{x=0,W}\, dx \qquad 5.6$$

where H and W are the height and the width of the domain. The meshes are denoted by 'Rnai' for i local refinement steps after n regular refinements. However, we have to explicitly state that the level of grid refinement towards the walls has not been chosen on

5.4. MIT BENCHMARK 2001

the basis of an a posteriori error indicator, but a priori only, since it was the primary aim of these studies to show that local alignment together with hanging nodes can be directly integrated into this monolithic approach without any loss of efficiency while gaining higher accuracy. The combination with user-defined a posteriori error control mechanisms which lead to an automatic grid refinement, resp., grid coarsening is part of future studies with this full Galerkin approach.

The level 2 mesh (R2, meaning '2 times regular refinement' of the coarse mesh) is used to perform the first computation until the solution reaches a periodical result (after 1500 non-dimensional time units), then the last output result is used as a starting point for the following computation. Note that regular refinement doubles the number of elements in both x and y-direction. The results of the MIT Benchmark 2001 configuration that are computed by our new approach oscillate periodically in time (Fig. 5.8) and are presented in Table 5.4. More results can be found in Appendix A.5.

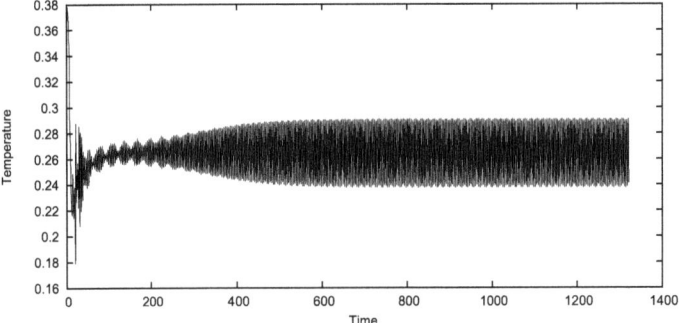

Figure 5.8: Temperature oscillations at point 1

Several comparisons have been made to see the differences among the other references. In [36] it is mentioned that the Q_2P_1 element with coarse mesh (27 x 121) performs poorly in the sense that the results show too low amplitudes for velocity and temperature at point 1 (0.00542 and 0.00442). In contrast, we observe good results even with the level 2 mesh (16 x 88). They also calculated Nusselt numbers that are slightly different from the reference, see Le Quéré [64]. In fact, we produce the same results (R2, R3, R4), but only as soon as we introduce local refinement near the wall, the Nusselt number improves strongly even with the level 2 mesh. It is obvious that the Nusselt number calculated on level 3 and 4 (R3 and R4) can be improved by using the level 2 mesh with local refinement (R2a1 and R2a5).

We believe that without local grid refinement we might have to use level 5 or higher to produce the same Nusselt numbers as the one produced by Le Quéré. This information shows us the expected result that local grid refinement helps a lot for this test configuration. The time step is not an issue as long as we put enough time steps over one period. 20 up to 40 time steps are already sufficient to produce excellent results for this problem, and no specific gain/loss in the quality of the Nusselt number has been observed if we increase/decrease the number of time steps in one period. Summarizing, the differences between our and the reference results are 0.02% for velocity u_1, 0.003% for temperature Θ_1 and only 0.004% for the Nusselt number.

CHAPTER 5. NUMERICAL VALIDATION

Table 5.4: Results of the MIT Benchmark 2001 simulations

Author	u_1	Θ_1	-Nu	Period
Turek	0.0572	0.2647	4.5791	3.422
Davis	0.0563	0.2655	4.5796	3.412
Gresho	0.05665	0.26547	4.5825	3.4259
Le Quéré	0.056356	0.26548	4.57946	3.4115
R2a0	0.058139	0.26539	4.66245	3.4000
R2a1	0.057674	0.26538	4.59295	3.4214
R2a5	0.057490	0.26540	4.57941	3.4214
R3a0	0.056787	0.26548	4.59318	3.4214
R3a1	0.056665	0.26546	4.58155	3.4214
R3a4	0.056591	0.26549	4.57967	3.4214
R4a0	0.056451	0.26549	4.58158	3.4200
R4a1	0.056394	0.26546	4.57994	3.4154
R4a3	0.056372	0.26546	4.57969	3.4214

5.5 Planar flow around cylinder

Planar flow around cylinder has been the benchmark to viscoelastic flow problems to test different material models and numerical methods. In this study, Oldroyd-B and Giesekus material models are implemented and applied within this configuration. The coarse mesh is depicted in Fig.5.9. The convective term inside the momentum equation is neglected (creeping flow) and the material parameters are set to $\beta = 0.59$ and $\eta_0 = 1.0$ for comparison with other references. Similar to [41, 1, 18], numerical results do not show pointwise convergence in the wake region for the Oldroyd-B model beyond Weissenberg number $\text{We} = 0.7$. Only for $\text{We} < 0.7$ our numerical results show a mesh independent behavior in a pointwise sense, but similar as in [41], we can show that the numerical stability increases significantly by using the LCR model and computations up to $\text{We} = 2.0$ or more are visible. In contrast to other approaches, the same numerical results are obtained by using a direct steady approach instead of time-dependent simulations. The full computational domain $(-20 < x < 20)$ has two symmetry lines which cross the center of the cylinder (origin $(0, 0)$) while the velocity inflow has one symmetry line coincide with the x-axis. It is therefore reasonable to consider only half of the domain for numerical simulation. The local refinements are applied around and in the wake of the cylinder to get more accurate results with less computational effort which is shown in the coarse mesh of Fig. 5.9. By local refinement we mean to refine for the next mesh level only those elements attached to the location that we prescribed before hand (bold lines in Fig. 5.9). One exemplary grid is shown in Fig. 5.10 for mesh R3a3. Our extensive tests for this configuration show that 3 times regular refinement plus additional local refinements is a very good compromise between accuracy and computational efforts. In the following, we show that the stress and the drag values for $\text{We} = 0.6$ lead to mesh independent results and are in a good agreement with [41] for mesh R3a4 and R3a5. The drag value converges, if local mesh refinement is used, which leads to a slightly higher number than in [41, 1, 27] but slightly lower than in [26], see Tab. 5.5. The stress plot depicts qualitatively the same one from [41].

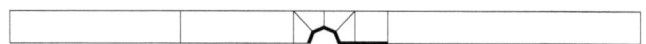

Figure 5.9: The coarse mesh for planar flow with a priori local refinement locations.

5.5. PLANAR FLOW AROUND CYLINDER

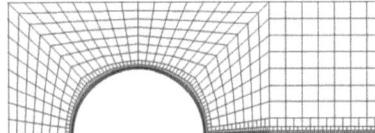

Mesh	NEL	DOF
R3a1	656	15823
R3a2	944	22457
R3a3	1520	35715
R3a4	2672	62221
R3a5	4976	115223
R3a6	9584	221217
R3a7	18800	433195

Figure 5.10: Computational mesh R3a3 with local refinement. Right: mesh information.

Table 5.5: **Oldroyd-B, We** = 0.6: NL denotes the number of nonlinear iteration. Peak 1 and 2 denote the peak value of stress on the cylinder and in the wake respectively. Initial solution is zero vector.

Level	R3a2	R3a3	R3a4	R3a5	Ref.[41, 27]	Ref.[26]
Drag	117.764	117.776	117.779	117.779	117.775	117.78
NL	10	10	10	10	-	-
Peak 1	95.591	94.443	94.354	94.294	-	-
Peak 2	15.877	16.738	17.247	17.377	-	-

Here, the Newton method behaves very stable with respect to different mesh levels which is shown by the number of nonlinear iterations. For all computed meshes, by using a direct steady approach, the Newton method needs 10 iterations starting from zero before satisfying the nonlinear stopping criterion that is set to 10^{-8}. Hence, the approximate solutions for We = 0.6 can be obtained with the same accuracy as other references results without having to use much computational efforts due to pseudo-time stepping. Fig. 5.11 clearly shows that mesh converged solution can be obtained for We = 0.6 with Oldroyd-B fluid. For different mesh levels, Peak 1 and Peak 2 seems to fully converge by mesh refinement. Although Peak 2 seems to converge later than Peak 1 by refinement, it does not show any mesh dependent solution. This is due to the fact that Peak 2 occurs in the wake region where high stress gradient is expected near the stagnation point of the cylinder.

Table 5.6: **Oldroyd-B, We** = 0.7: Newton behavior and dimensionless drag values. Initial solution is zero vector. Reference value of drag is 117.323 [1].

Level	R3a2	R3a3	R3a4	R3a5	R3a6	R3a7
Drag	117.302	117.316	117.321	117.322	117.323	117.323
NL	11	11	11	11	11	11
Peak 1	108.466	107.061	107.246	107.104	107.055	107.068
Peak 2	27.539	31.167	34.993	37.809	41.189	43.086

The presented methodology also recovers the same problem as other authors when computing for We = 0.7. Here, the first normal stress converges in a pointwise sense only around the cylinder region to that value which is denoted as Peak 1, while in the wake region it does not show mesh independent solutions with increasing the local refinement. As already predicted from computation of We = 0.6, the mesh converged solution is getting difficult for We = 0.7. In this configuration, Oldroyd-B fluid model suffers in the high stretching region which is similar to extensional flow[1], see Appendix A.4. Nevertheless, the drag (contains not

---
[1] Extensional flow describes uniaxially stretching the fluid in both directions. Filament stretching is one example of such a flow

CHAPTER 5. NUMERICAL VALIDATION

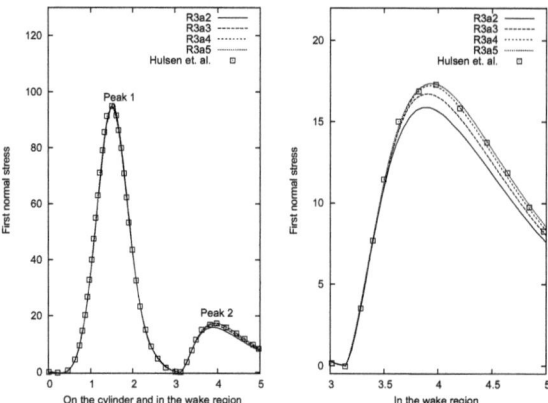

Figure 5.11: **Oldroyd-B, We = 0.6:** Normal stress convergence with local refinement. Right: zoom of the wake part.

only stress but also pressure values) converges to a value which is the same as in [1], see Tab. 5.6.

Table 5.7: **LCR, Oldroyd-B, mesh R3a4:** Newton behavior and dimensionless drag values. NL denotes the number of nonlinear iterations.

We	Drag	NL	We	Drag	NL	We	Drag	NL
0.1	130.366	8	0.8	117.347	4	1.5	125.665	4
0.2	126.628	5	0.9	117.762	4	1.6	127.523	4
0.3	123.194	4	1.0	118.574	6	1.7	129.494	4
0.4	120.593	4	1.1	119.657	6	1.8	131.578	4
0.5	118.828	4	1.2	120.919	5	1.9	133.754	4
0.6	117.779	4	1.3	122.350	4	2.0	136.039	5
0.7	117.321	4	1.4	123.936	4	2.1	138.438	5

The same as before, for We = 0.7, the Newton method needs a number of iterations which is independent of the used mesh levels. Thus, for critical We number in the case of Oldroyd-B model, the approach remains stable, too.

The advantage of our approach gets clear from the two examples of We = 0.6 and We = 0.7. Although the results for We = 0.7 may not necessarily represent the exact solution in the wake for Oldroyd-B (if any), the approach allows numerically stable solutions for higher values of We, see Tab. 5.7. Here, the Newton method needs only few iterations to satisfy the stopping criterion which we set to 10^{-8} when increasing the We number continuously by 0.1. For this table, the initial solution is zero for We = 0.1 while for the other We numbers, the initial solution uses the solution of the previous We number. The advantage of EO-FEM is very pronounced in this case, namely by stabilizing the nonlinear term inside LCR. Thus, numerical computations for higher We numbers are made to be possible.

On the other hand, the viscoelastic model with conformation stress tensor formulation for

5.5. PLANAR FLOW AROUND CYLINDER

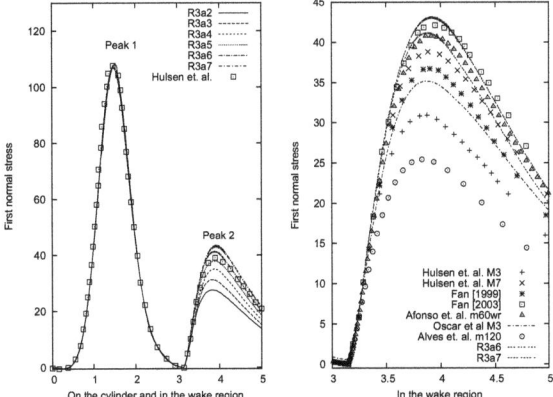

Figure 5.12: **Oldroyd-B**, We = 0.7: Mesh dependent solutions from several authors. Right: zoom of the wake part.

Table 5.8: **Conf., Oldroyd-B, mesh R3a4:** Newton behavior and dimensionless drag values. NL denotes the number of nonlinear iterations. The initial solution uses lower level for We = 0.1 while for the other We numbers, the initial solution uses the solution of the previous We number.

We	Drag	NL	We	Drag	NL	We	Drag	NL
0.1	130.366	2	0.4	120.598	3	0.7	117.344	4
0.2	126.629	3	0.5	118.837	4	0.8	117.392	5
0.3	123.196	3	0.6	117.792	4	0.9	div.	35

Oldroyd-B, equation (2.9), is not able to converge at We = 0.9, see Tab. 5.8. The numerical divergence can be due to direct steady approach while the solutions might be already unsteady at We = 0.9, or it can be due to the model itself which already produces mesh dependent solution at We = 0.7. While the latter argument is a fact, there is no unsteady flow behavior can be observed even with full domain numerical experiment to support the first argument. Hence, the LCR model is very potential in this configuration.

The plotted results of Tab. 5.7 are very comparable to what other authors have presented, see Fig. 5.13. Our approach recovers what other numerical schemes with LCR are able to do, however, with a more efficient solution method.

At this point, we also show the influence of the nonlinear term inside the momentum equation onto the drag and stress values on the cylinder and in the wake region for We = 0.6 and We = 0.7. Here, the Oldroyd-B model is coupled with the nonlinear Navier-Stokes equation and the same inflow parameters as the ones for the creeping flow are set. This setting gives a small amount of inertia effect which leads to $Re = 1.0$. Tab.5.9 shows that by retaining the nonlinear term $(\mathbf{u} \cdot \nabla \mathbf{u})$, the peak values of the stress are less than if we neglect the term (Stokes flow), compare with Tables 5.5 and 5.6. The drag values, which are converged for both We = 0.6 and We = 0.7, are bigger than if we neglect the nonlinear term.

CHAPTER 5. NUMERICAL VALIDATION

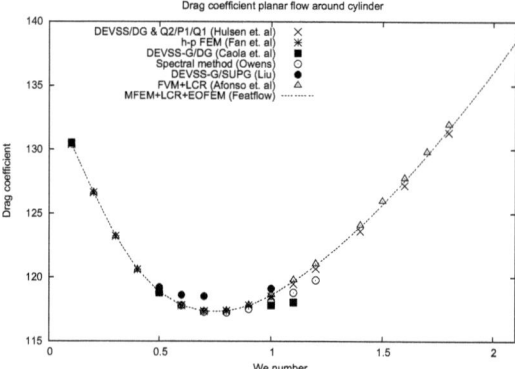

Figure 5.13: Non dimensional drag values from different authors.

It means that by introducing the inertia, the elasticity effect reduces accordingly and vice versa. Thus, the drag becomes bigger. Numerically, by introducing the inertia, the Newton method needs more iterations to satisfy the same stopping criterion. But apart from that, our numerical approach is stable whether the nonlinear term in the momentum equation is neglected or not.

Table 5.9: **Oldroyd-B with Navier-Stokes:** Newton behavior and dimensionless drag values for We = 0.6 and We = 0.7. Initial solution is zero vector.

Level	R3a2	R3a3	R3a4	R3a5	R3a6	R3a7
We = 0.6						
Drag	118.526	118.539	118.543	118.544	118.544	118.544
NL	13	13	13	13	13	13
Peak 1	91.978	91.831	92.137	92.007	91.971	91.984
Peak 2	12.802	13.191	13.473	13.559	13.643	13.727
We = 0.7						
Drag	118.218	118.230	118.234	118.235	118.236	118.236
NL	14	14	14	14	14	14
Peak 1	104.067	103.670	104.110	104.018	103.986	103.985
Peak 2	22.163	24.252	26.194	27.498	28.779	29.541

Next, in the same way as in [41], we validate the same approach by performing simulations for the Giesekus model, with the specific choice of mobility factor $\alpha = 0.01$. The model has a better pointwise mesh convergence in the wake than in the case of the Oldroyd-B model due to the fact that Giesekus model introduces a nonlinear term to control extensional flow in the wake region, see Appendix A.4. It is shown in Fig. 5.14 that our approach leads to the same result of cutline of the conformation stress tensor as in [41] for We = 5 with the direct steady approach. Furthermore, we perform the same simulation for We = 5 with the corresponding non steady approach, which confirm the same results of Peak 2 and drag values.

5.5. PLANAR FLOW AROUND CYLINDER

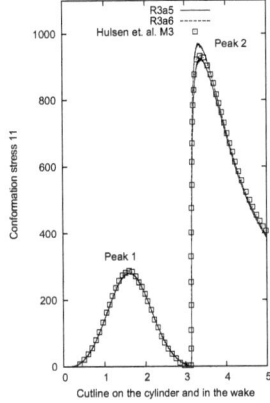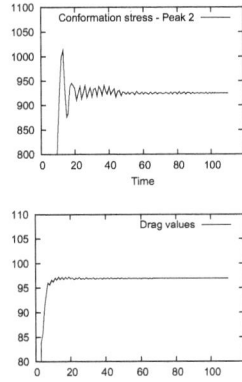

Figure 5.14: **Giesekus model, We** = 5: Left: direct steady approach resulting in drag= 96.9429. Right: non steady approach with mesh R3a5 only.

Table 5.10: **Giesekus model, mesh R3a5:** Newton behavior and dimensionless drag values. NL denotes the number of nonlinear iterations and Peak 2 denotes the second peak of the conformation tensor. The initial solution is zero for **We** = 5 while for larger **We** numbers, the initial solution is the solution of the previous **We** number.

We	Drag	Peak 2	NL	We	Drag	Peak 2	NL
5.0	96.943	924.45	14	60	85.859	12010.57	4
20	89.905	4204.51	12	70	85.356	13773.61	4
30	88.304	6318.79	5	80	84.937	15502.45	4
40	87.256	8311.32	5	90	84.585	17207.87	4
50	86.476	10199.10	4	100	84.287	18897.95	4

The computation for **We** = 20 uses the initial solution of **We** = 5 and we consecutively compute higher **We** numbers up to **We** = 100 by using mesh R3a5, see Tab. 5.10. From this table, we can see again that the approach remains stable and needs only few iterations before satisfying the nonlinear stopping criterion which we set to 10^{-8}. This shows that the solver is very strong, i.e. the Jacobian at every nonlinear step gives a very good approximation and thus a good direction for the next iteration irrespective of the chosen viscoelastic models. The treatment of Jacobian by divided difference approach can really be a 'black-box' for any other viscoelastic models in the coming future. The corresponding non steady approach for **We** = 100 with the same mesh R3a5 can be seen in Fig. 5.15 which again leads to the same results for the peak 2 and drag value as the one with the direct steady approach.

CHAPTER 5. NUMERICAL VALIDATION

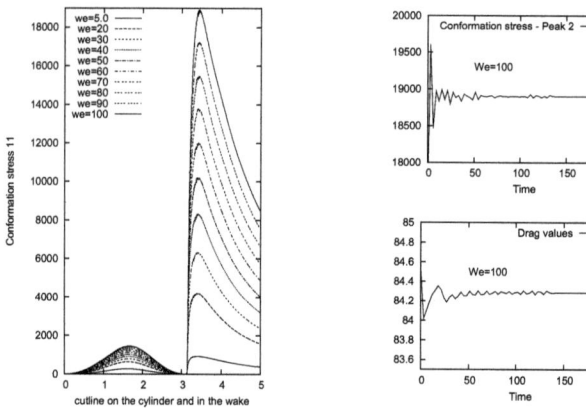

Figure 5.15: Giesekus model with $\alpha = 0.01$. Left: direct steady approach for We $= 5, 20, 30, 40, 50, 60, 70, 80, 90, 100$. Right: non steady approach for We $= 100$.

Applications

This chapter is dedicated to the application of 2D flow simulations that show interesting physical phenomena. Basically, they are separated in two sections, namely nonisothermal and viscoelastic flow. These simulations include the flow in a heat exchanger, the lip-vortex growth in a contraction viscoelastic flow and viscoelastic flow in a cavity. These complex flows behavior are of industrial and research interest.

6.1 Nonisothermal flow

In this section, two examples of temperature dependent viscosity flows and one example of heat dissipation flow are presented. The first two examples have different configurations for different industrial purposes by setting $k_2 = 0$ in equation (2.2), while the last example (by setting $k_2 = 1$ in the same equation) has a configuration that may be used in the viscoelastic section of contraction flow.

6.1.1 Temperature dependent viscosity: heat exchanger

Nonisothermal flows can be found in any heat exchanger appliances. The main idea of this tool is to transfer heat through channels with Newtonian fluid as media transport. Hot water comes into the channels, and heat is released through channels wall. In order to demonstrate the flexibility and efficiency of the corresponding numerical experiments, the flow is controlled by temperature-dependent viscosity relation. It means that a slightly lower temperature value as Dirichlet boundary data is set at one of the channel. Thus, a localized larger viscosity value will qualitatively 'stop' the flow on the corresponding channel. Prototypical test simulations are set with a given viscosity function

$$\eta = \eta_0 \, e^{\left(a_1 + \frac{a_2}{a_3 + \Theta}\right)} (b_1 + b_2 \, \|\mathbf{D}\|)^{-b_3} \qquad 6.1$$

with specific parameters $a_1 = a_2 = 1.0, a_3 = 179.4, b_3 = 0, \eta_o = 1.0$. Here, $\|\mathbf{D}\| = \sqrt{D_{ij}\,D_{ij}}$ is the magnitude of the usual symmetric part of the gradient velocity, $\mathbf{D} = \frac{1}{2}(\nabla \mathbf{u} + \nabla \mathbf{u}^T)$. The

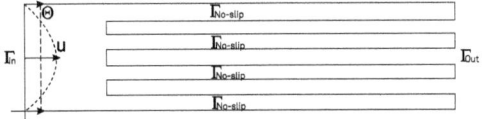

Figure 6.1: Problem set up.

CHAPTER 6. APPLICATIONS

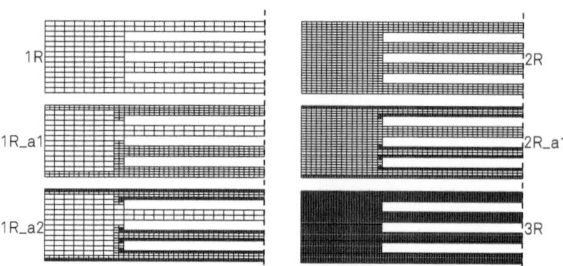

Figure 6.2: Computational mesh with local refinement.

Table 6.1: Mesh informations

Mesh	Cells	Nodes	Edges	Dof
R1a0	752	1095	1846	13335
R1a1	2264	2849	5112	37467
R1a2	5264	5847	11110	82455
R2a0	3008	3693	6700	49227
R2a1	6008	6691	12698	94215
R3a0	12032	13401	25432	188691

chosen material parameters do not yet correspond to any certain type of fluid. The specific geometry (width = 3.5 and length = 44 in non-dimensional units) and setting are given in Fig. 6.1, which shows 4 channels installed for low Reynold number flow (Re $\approx$ 14). The hot fluid enters the inflow section with non-dimensional temperature $\Theta = 250$ and a parabolic profile of velocity. Heat is then distributed to all channels by the diffusivity of the heat transfer model in equation (2.2). We specify Dirichlet data for the temperature ($\Theta = 190$) to all channels except to the second channel ($\Theta = 180$) to control the fluid flow at this pipe through the viscosity function. On this geometry, we compute for mesh R1a0, R2a0, R3a0, R1a1, R1a2 and R2a1 (see Fig. 6.2). The local refinement is set for channel 1, 3, and 4. All initial solutions start from zero. Table 6.2 shows how the proposed method converges

Table 6.2: NL/MG for different levels, with Tol denoting the linear stopping criterion

Tol/Level	R1a0	R2a0	R3a0	R1a1	R1a2	R2a1
10^{-1}	10/1	9/1	10/1	9/1	9/1	8/1
10^{-2}	7/1	7/2	7/2	7/1	7/1	7/2
10^{-3}	6/1	6/2	6/3	6/1	6/1	6/2
10^{-8}	5/5	5/6	5/8	5/4	5/4	5/5
UMFPACK	5/-	5/-	4/-	5/-	5/-	5/-

with respect to a given number of digits for the linear multigrid solver. Here, NL denotes the number of Newton iteration, while MG presents the averaged number of multigrid iterations per nonlinear step. We can see from Table 6.2 that the local refinement does not disturb the multigrid convergence. The memory requirement of the used computer is still capable to run

6.1. NONISOTHERMAL FLOW

a direct linear solver (UMFPACK) for this problem at least up to 3 refinements. In this case, the multigrid performance is getting closer to UMFPACK (with respect to the number of nonlinear iterations) by decreasing the linear tolerance. It is clear that for a linear tolerance value of TOL= 10^{-8} the multigrid needs more or less the same number of nonlinear iterations as UMFPACK.

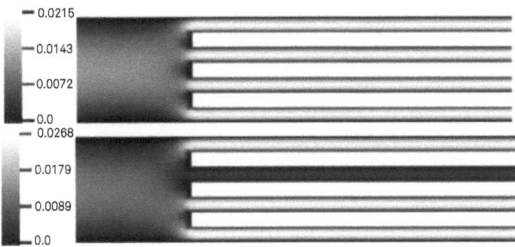

Figure 6.3: The Euclidean norm of velocity. Top: The flow is not blocked. Bottom: The flow at the second channel is blocked

As it is already expected, this Dirichlet temperature difference will nevertheless increase the local viscosity and hence 'stop' the flow at the corresponding channel. Fig. 6.3 shows the resulting flow behavior which almost stops at the second channel caused by a locally growing viscosity. In the left figure, the same Dirichlet temperature ($\Theta = 190$) is prescribed to all channels (the viscosity value is set to 0.001 at all channels) while in the right figure, the Dirichlet temperature at the second channel is set to $\Theta = 180$. Hence, viscosity at this channel grows to 0.0144 ($\approx$ 14 times bigger than before), and finally the flow is 'stopped' at this channel.

6.1.2 Temperature dependent viscosity: micro-reactor

In the following, the simulation is extended with respect to a more complex geometry, which has many channels inside. This configuration is prototypically used in chemical micro-reactor processes of heat and mass transfer to gain more efficiency. Complex geometry could be

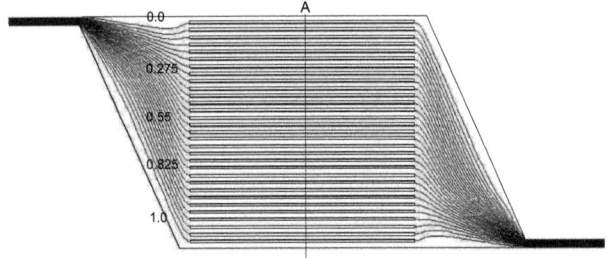

Figure 6.4: Streamlines for flow with constant viscosity

problematic in numerical computations due to its aspect ratio which is not convenient in most cases. Finite element method is well-known also for dealing with wide range of geometric

CHAPTER 6. APPLICATIONS

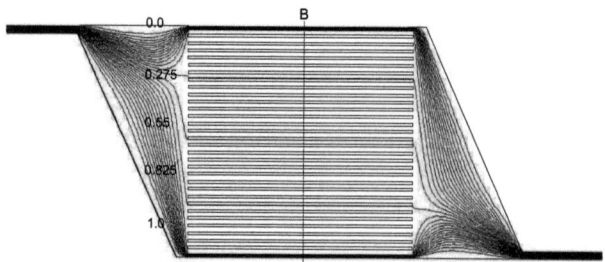

Figure 6.5: Streamlines for flow with nonlinear viscosity

complexity. Hence, it is a good example on this work to see how the solver may behave for such a geometry by taking into account additional heat transfer equation and nonlinear viscosity function. Fig. 6.4 shows typical flow profiles of flow simulation with constant viscosity ($a_1 = a_2 = b_3 = 0$), while in Fig. 6.5 shows the corresponding flow simulation with nonlinear viscosity. Here, the flow of all middle channels is blocked, again by prescribing different wall temperature values as boundary condition. Thus, the flow goes only through the top and the bottom channels. Fig. 6.6 shows clearly the different flow profiles with respect to the different viscosity parameters. The flow in the middle channels, which is almost zero, may cause numerical breakdown up to some extend. In this work, simulations for both settings can be done without much difficulty. The first two examples of heat exchanger leads to the

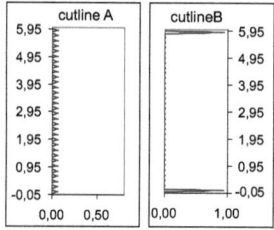

Figure 6.6: Cutline A and B of the velocity magnitude

idea of a 'non-mechanical valve' which can be controlled by setting the outside temperature only. This kind of flow settings can be numerically done within monolithic approach in a strong manner. As a prototypical application, one may think of hot pattex material (for glue purposes), which can flow when the temperature increases and which will become an elastic solid when the temperature decreases.

6.1.3 Heat dissipation

In this subsection, the effect of adding a viscous dissipation term into the equation of energy is analyzed. The additional dissipation term, which is just the symmetric part of the velocity gradient, is activated by setting $k_2 = 1$ in equation (2.2). This additional term can be physically viewed as viscous heat generation along with the fluid flow. The heat, which is generated from this friction, may dramatically change the temperature and velocity profile of

6.1. NONISOTHERMAL FLOW

the flow. Thus, it can be interesting for polymer flow modelling.

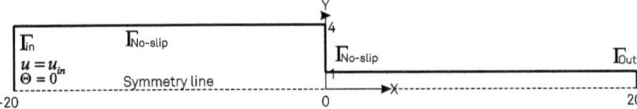

Figure 6.7: Geometry of the 4 to 1 contraction configuration

The heat transfer model is tested with constant viscosity ($\eta = \eta_0$) for the well-known 4 to 1 contraction geometry. A non-slip boundary condition at the upper wall is set, while a Dirichlet condition is prescribed at the inflow (see Fig. 6.7). By neglecting the time derivative,

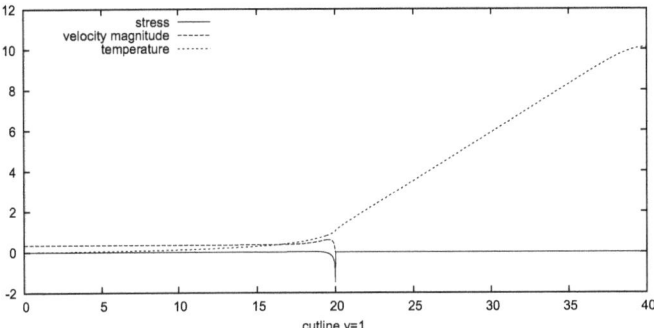

Figure 6.8: The cutlines of $T_{11}, \theta,$ and u_x

$\dfrac{\partial \Theta}{\partial t} = 0$, one directly calculates the stationary result with the proposed Newton-multigrid solver. Although zero temperature is prescribed at the inflow, heat is produced along the channel as the friction becomes higher. Generally, it gives additional heat locally as the material begins to flow. Fig. 6.8 shows the qualitative effect of the dissipation term. One can see from the cutline diagram ($y = 1$) that near the inflow ($-20 < x < 0$), heat is produced slowly. Yet after entering the contraction ($0 < x < 20$), there is a big gradient of temperature

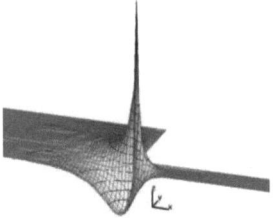

Figure 6.9: 3D representation of stress tensor component for the $2R\_a1$ mesh

CHAPTER 6. APPLICATIONS

up to the end of the channel. The components of velocity and of the stress tensor also show that the computation may face numerical difficulties around the entrance. The characteristic of such a geometry appears clearly from the entrance point where the sharp corner point may cause numerical problems in viscoelastic flow since then the stress tensor will raise to a huge value (see Fig. 6.9). So, local refinement around the entrance corner and also at the end of the channel are very important to get a better capture of the temperature field.

Table 6.3: NL/MG for different levels, with Tol denoting the linear stopping criterion

Tol/Level	R1a0	R2a0	R3a0	R1a1	R1a2	R2a1
10^{-1}	11/1	12/2	12/2	13/1	12/1	11/1
10^{-2}	9/2	10/3	10/4	9/2	9/2	9/2
10^{-3}	9/4	9/5	9/6	9/2	9/3	9/3
10^{-8}	9/13	9/17	9/19	9/7	9/7	9/9
UMFPACK	9/-	9/-	9/-	9/-	9/-	9/-

Tab. 6.3 shows how the Newton-Multigrid and Newton-UMFPACK solver converge for this kind of problem. It is clear that the proposed method seems to be robust with respect to the use of local refinement. As a test configuration, a low inertia flow is set, Re ~ 2.5. This is typically controlled by the velocity profile, characteristic length of the width of channel, and the given viscosity value. The meshes as well as the isoline of axial stress tensor are shown in Fig. 6.10, which stresses the role of adaptive refinement. Here, mesh $1R\_a1$ is comparable with mesh $2R$, and mesh $1R\_a2$ or mesh $2r\_a1$ are comparable with mesh $3R$. This situation is also illustrated in Fig. 6.11 where the axial stress profiles are compared along the entrance line ($x = 0$).

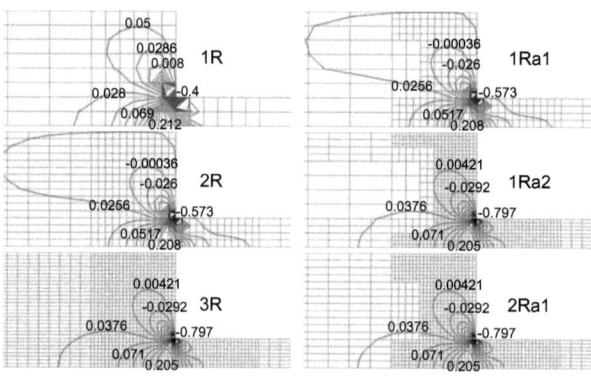

Figure 6.10: Isolines for the stress tensor on several different meshes. Note the different scale between mesh R1a0, R2a0, and R3a0

These prototypical studies have to be seen as preparing simulations for more complex flow configurations for which the extra stress tensor **T** will be extended to a model that adds elastic property in the case of viscoelastic fluid flow. This gives a better physical meaning of additional heat along the flow due to viscous dissipation and makes an interesting subject for the numerical tests.

6.2. VISCOELASTIC-RELATED FLOW

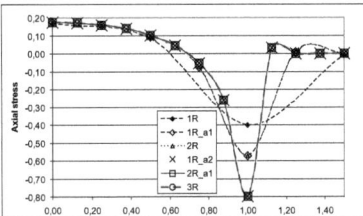

Figure 6.11: Axial stress profile along the entrance line $x = 0$

6.2 Viscoelastic-related flow

6.2.1 Lip vortex growth

The 4:1 Contraction problem is one of the most well-known benchmarks for viscoelastic flow. The difficulties of this configuration are due to the sudden contraction from width 4 units to 1 unit, hence causing an extensional flow at the downstream channel and a recirculation zone at the corner. The aim of this section is to describe the growth mechanism of the lip vortex as shown in some experiments by Boger and Walter, see [11], where its enhancement at least depends on the type of contraction, the flow rate and the properties of the fluid. Similar to the planar flow around cylinder configuration, we prescribe Dirichlet data for velocity at the inflow which is parabolic and we set the outflow velocity to natural conditions [37] and no-slip condition at the solid walls. The stress inflow boundary is computed in the same way as before. The outflow and solid wall stress boundaries are set to natural boundary conditions. Here, we apply a priori local refinement around the corner to produce a smooth streamline for

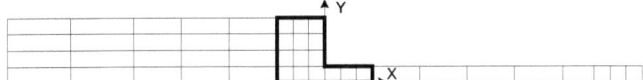

Figure 6.12: Coarse mesh and local refinement location.

Figure 6.13: Computational mesh for the contraction problem.

capturing the lip vortex as can be seen from Fig. 6.12 showing the coarse mesh. A coupled finite element approach with a Newton solver was presented for viscoelastic flow in [51], where they used a streamline-upwind method and found a weak salient corner in size. The lip vortex appears above We $= 2.0$ but disappears as the We number increases. Furthermore, in [49, 50], streamline-upwind does not discover the lip vortex in the finite element context. In contrast to those, finite volume with a staggered grid approach has found the increasing lip vortex in size as We number increases, see [76]. Hence, lip vortex enhancement depends on the numerical techniques, the stabilization method and the mesh size.

In this study, as already mentioned, we use the fully coupled approach with consistent

CHAPTER 6. APPLICATIONS

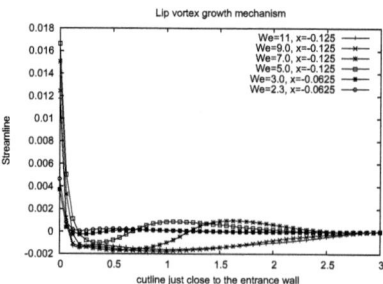

Figure 6.14: Cutlines of stream function at $x = -0.0625$ and $x = -0.125$.

edge-oriented stabilization based on LCR for Oldroyd-B type of fluid. The mesh is shown in Fig. 6.13 where the distribution of elements is concentrated around the entrance corner utilizing hanging nodes. The lip vortex starts to appear at We = 2.3 that, however, is hardly visible in our case. Fig. 6.14 shows cutlines for the stream function close to the entrance wall where we can see the passage of the lip vortex as the cutline crosses the zero line for We = 2.3. Here, we take a line as close as possible to the wall in order to detect the first time appearance of the lip vortex as the We number increases. The width from the y-axis to this line can be taken as the smallest width of the cell close to the lip entrance, which in our case, corresponds to the width of the smallest element after refinement, $h = 0.0625$. As the lip vortex begins to grow, one may shift this line one cell width to the left in order to cross the lower curvature of the stream function. Fig. 6.15 illustrates the appearance, the growth and the collision of the lip vortex and the salient corner vortex as the We number grows from We = 1.0 to We = 11.0 that is accompanied by a decreasing corner vortex. The appearance

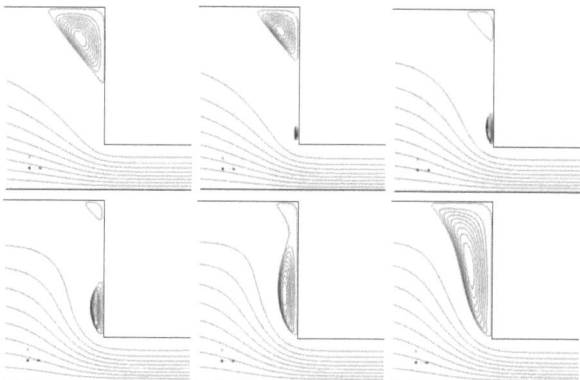

Figure 6.15: The growth of lip vortex. First row: We = 1.0, 3.0, 5.0, second row: We = 7.0, 9.0, 11.0.

of the lip vortex, which is above We = 2.0, is in a good agreement qualitatively with previous

6.2. VISCOELASTIC-RELATED FLOW

studies. Here, one can clearly see the streamline separation between the lip and the salient corner vortex before both vortices join together at We $= 9.0$ and then form one big vortex at We $= 11.0$.

6.2.2 Viscoelastic in a cavity

We consider the numerical simulation of both direct steady and nonstationary flow in a lid-driven cavity for the Oldroyd-B model. The initial condition for the stress tensor is unity and a regularized velocity boundary condition is implemented such that $\mathbf{u}(\mathbf{x},t) = (8(1+\tanh 8(t-0.5))x^2(1-x)^2, 0)^T$ on the top boundary while zero velocity on the rest of boundary is prescribed. For direct steady simulations, the velocity profile evolves to $\mathbf{u}(\mathbf{x},t) = (16x^2(1-x)^2, 0)^T$ on the boundary. The total viscosity (zero-shear viscosity) is set to $\eta_0 = 1$. The simulation is performed with the mesh size $h = 1/64$ and with coarse mesh size $h = 1/4$. The time step is chosen to be $\Delta t = 0.1$ in the sense that no further improvement in kinetic energy with respect to smaller time steps could be observed. The number of cells for the corresponding computation level n is $\mathrm{L}n = 2^{4+2n}$. We calculate the kinetic energy by $\frac{1}{2}\|\mathbf{u}_h\|_{0,\Omega}^2 dx$ and analyze the impact of jump stabilization for different We numbers. For We $= 1$, the kinetic energy

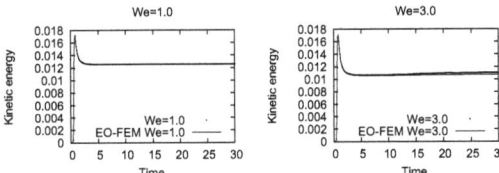

Figure 6.16: Kinetic energy until $t = 30$ for different We numbers with and without EO-FEM.

seems to reach a steady state as shown in Fig. 6.16 and it remains steady at least up to time $t = 30$. For We number bigger than 1, there is a tendency that the solutions start to

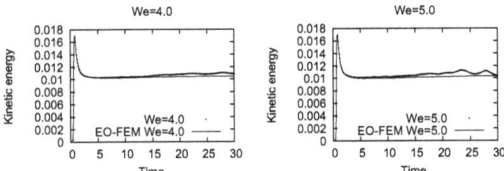

Figure 6.17: Kinetic energy until $t = 30$ for different We numbers with and without EO-FEM.

be nonstationary at longer time computation, $t > 30$. At the same time, the cutline of LCR component starts to oscillate for bigger We number, which later on may break the numerical computation. Yet, as the We number increases, the LCR variable does not show spuriousity at time $t = 30$ if EO-FEM is applied, see Fig. 6.18. At even bigger We number on the same time, $t = 30$, oscillations become clear and visible for LCR variable without EO-FEM. On the contrary, EO-FEM stabilizes the oscillations at least up to time, $t = 30$, which may avoid numerical breakdown at longer time $t > 30$, see Fig. 6.19.

Furthermore, it is interesting to see the streamline visualization of the flow inside of the domain as shown in Fig. 6.20. The same phenomena as in high inertia flow, we observe also a change of the streamline patterns as the We number increases. At lower We numbers, the

CHAPTER 6. APPLICATIONS

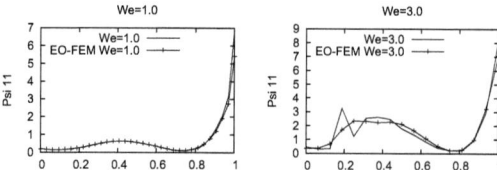

Figure 6.18: Cutline of ψ_{11} at $x = 0.5$, $t = 30$, for different We numbers with and without EO-FEM.

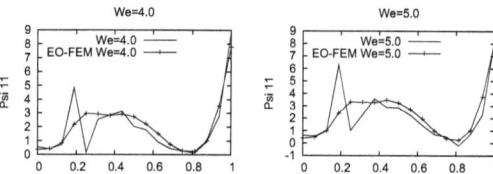

Figure 6.19: Kinetic energy until $t = 30$ for different We numbers with and without EO-FEM.

bottom corner vortex looks equally the same. Yet, as the We number increases, the lower left vortex decreases while the right one evolves in size. Thus, it shifts the whole flow field to the left.

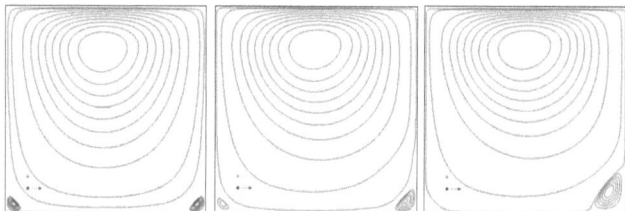

Figure 6.20: Shape changing of corner vortex in a cavity flow. From left to right We $= 1, 3, 4$.

6.2.3 Non-isothermal viscoelastic flow in 4:1 contraction

The following example is simulation of the full system of equation (6.2), which is a coupling of equation (2.17) and energy equation, in 4 to 1 contraction configuration.

$$\begin{cases} \rho\dfrac{\partial \mathbf{u}}{\partial t} + \rho(\mathbf{u}\cdot\nabla)\mathbf{u} = -\nabla p + \eta_s\Delta\mathbf{u} + \dfrac{\eta_p}{\Lambda}\nabla\cdot e^{\psi} + \rho(1-\gamma\Theta)\mathbf{j} \\ \qquad\qquad\qquad\qquad\qquad\qquad \nabla\cdot\mathbf{u} = 0 \\ \dfrac{\partial \psi}{\partial t} + (\mathbf{u}\cdot\nabla)\psi - (\Omega\psi - \psi\Omega) - 2\mathbf{B} = \dfrac{1}{\Lambda}\left(e^{-\psi} - \mathbf{I}\right) \\ \dfrac{\partial \Theta}{\partial t} + (\nabla\Theta)\mathbf{u} = k_1\,\nabla^2\Theta + k_2\,exp(\psi):\mathbf{D} \end{cases} \qquad 6.2$$

6.2. VISCOELASTIC-RELATED FLOW

The computational domain and inflow settings are the same as in the previous subsection 6.2.1. The viscous dissipation of the temperature field is integrated by replacing $k_2 \mathbf{D} : \mathbf{D}$ with $k_2 \exp(\psi) : \mathbf{D}$ to have a more physical meaning. After dropping all time derivative term, $\frac{\partial}{\partial t}$, a steady state solution is sought for We $= 2.0$. According to previous subsection 6.2.1, We $= 2$ is where a lip-vortex about to appear at the entrance. Here, the inflow velocity is set to parabolic profile in x-direction and let the outflow velocity to be Neumann boundary condition. No-slip condition is applied as usual at solid walls. The inflow stress profile is pre-computed from the given velocity profile, and let the rest stress boundary to be Neumann. The inflow temperature is set to be zero, so there is no heat coming into the domain. But as the flow begins, heat is generated along the flow field due to the viscous dissipation term inside the energy equation, $\exp(\psi) : \mathbf{D}$. And the generated heat will change the viscosity of the fluid as well through the relation:

$$\eta = \eta_0 \, e^{\left(a_1 + \frac{a_2}{a_3 + \Theta}\right)} (b_1 + b_2 \, \|\mathbf{D}\|)^{-b_3} \qquad 6.3$$

where a_1, a_2, a_3 and b_1, b_2, b_3 are specific material parameters. The increasing temperature

Figure 6.21: Temperature field over the domain

can be seen in the above Fig. 6.21 where it increases along the solid wall. In the contraction zone heat is further generated and reaches its highest value at the downstream solid wall. The viscosity field changes also and is seen in Fig. 6.22 where it decreases its value at the

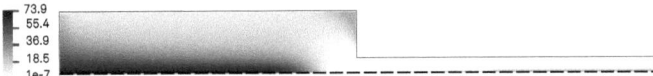

Figure 6.22: Temperature and shear dependent viscosity field over the domain

downstream. This is partly due to the increasing temperature. For this simulation we set the material parameters as $a_1 = a_2 = a_3 = 1.0, b_1 = b_2 = 1.0, b_3 = 10$ with thermal diffusivity $k_1 = 0.001$ and $k_2 = 1.0$.

6.2.4 Future extension

The flexibility of monolithic approach is quite visible from the given examples of non-isothermal and viscoelastic-related applications and experiments. The monolithic approach can be extended as well into other interesting application such as multiphase viscoelastic fluid flows. In the following, we propose one extension of an application in the area of biodynamics, especially in the medical image-based simulation which mainly deals with flowing blood particles in arteries, see [42]. The set of equations to solve are separated in two regions, namely the solid ($\phi_s < 0$) and fluid ($\phi_s >= 0$) region. Inside the solid, where the particle behaves elastic, an Oldroyd derivative of the Cauchy-Green deformation tensor is being used. Having no deformation as numerical variable, an ALE approach can be avoided. Thus the whole system

CHAPTER 6. APPLICATIONS

can be solved in a full Eulerian monolithic approach.

$$\begin{cases} \rho\dfrac{\partial \mathbf{u}}{\partial t} + \rho(\mathbf{u}\cdot\nabla)\mathbf{u} = -\nabla p + 2\eta\nabla\cdot\mathbf{D} + H(\text{sign}(\phi_s))G\nabla\cdot\mathbf{B}' \\ \nabla\cdot\mathbf{u} = 0 \\ \dfrac{\partial \phi_s}{\partial t} + (\mathbf{u}\cdot\nabla)\phi_s = 0 \\ \dfrac{\partial \mathbf{B}}{\partial t} + H(\text{sign}(\phi_s))\left[\mathbf{u}\cdot\nabla\mathbf{B} - \nabla\mathbf{u}\cdot\mathbf{B} - \mathbf{B}\cdot\nabla\mathbf{u}^T\right] + (1 - H(\text{sign}(\phi_s)))\mathbf{B} = 0 \end{cases} \quad 6.4$$

with a heavy-side function

$$H(\text{sign}(\phi_s)) = \begin{cases} 1 & \text{for } \phi_s < 0 \text{ (solid region)} \\ 0 & \text{for } \phi_s >= 0 \text{ (fluid region)}. \end{cases} \quad 6.5$$

Here, G is the modulus of transverse elasticity, $\mathbf{B}$ is the left Cauchy-Green deformation tensor. The solid stress can be based on: neo-Hookean material, The Mooney-Rivlin model or the Saint Venant-Kirchhoff model. The deviatoric stress is defined as $[.]' = [.] - \frac{1}{2}(\text{tr}[.])\,\mathbf{I}$ for 2D. The third equation of equation (6.2.4) is the level set equation, which is a pure hyperbolic transport problem. This interface tracking method is easily used by just using the sign of ϕ_s to identify the different domain (fluid-solid). On the other hand, the oldroyd derivative of the left Cauchy-Green deformation tensor may cause a numerical instability. Thus, a numerical stabilization technique is needed for both cases, which in this study by using a jump stabilization technique, EO-FEM.

Reversibility shape

To give a quick short that the monolithic approach may deal with the above set of equations, we test a reversibility shape. A solid circle is set in the middle of a square domain [-1,1] with radius $r = 0.75$ while given a shear at the top and bottom with different direction. Thus, the solid will be stretched. The density is equal for solid and fluid ($\rho = 1$), and there is no body force acting. The viscosity of the fluid is set to $\mu = 1.0$ with constant $G = 4.0$. External forces are applied by setting Dirichlet boundary for x-velocity to be 1(at the top) and -1(at the bottom). These external forces are applied up to time $t = 0.7$ and then released. They

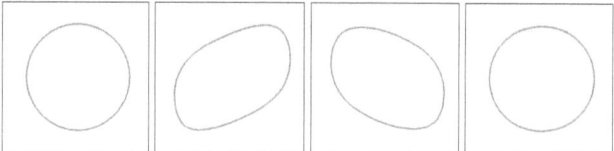

Figure 6.23: Circularity at time $t = 0, 0.7, 9.1, 15$.

are applied again between time $3.7 < t < 4.4$ and then released. At time $t = 7.4$ the forces are applied one more time but in different direction i.e, -1(at the top) and 1(at the bottom) and then released at time $t = 8.1$ up to the end, see Fig. 6.23. The initial solution is found by considering everywhere fluid region and solving a direct steady set of the following equations:

$$\begin{cases} \rho(\mathbf{u}\cdot\nabla)\mathbf{u} = -\nabla p + 2\eta\nabla\cdot\mathbf{D} \\ \nabla\cdot\mathbf{u} = 0 \\ \phi_s = x^2 + y^2 - r^2 \\ \mathbf{B} = \mathbf{I} \end{cases} \quad 6.6$$

6.2. VISCOELASTIC-RELATED FLOW

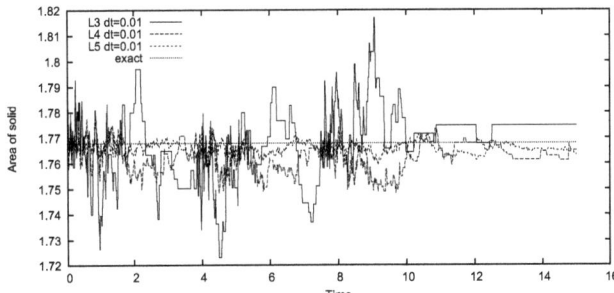

Figure 6.24: Area plot over time.

with **B** is propotional to identity so that the deviatoric stress is zero at rest, and r is the radius of the initial solid. The plot of the solid area over the time when the external forces are applied is shown in Fig. 6.24. We can see that by increasing refinement the area plot is getting closer to the exact value.

7
Summary and Outlook

This work has been about development, analysis and implementation of numerical methods for solving both nonisothermal and viscoelastic fluid flows of Oldroyd-B and Giesekus type. The aim of this work is to gain more insight in the issue of solving numerically viscoelastic flow simulations.

The work begins by introducing the physical set of equations, which couple the generalized Navier-Stokes and the stress equation. The reformulation of the stress equation is presented and given in the LCR approach. This approach is able to capture high stress gradients that occur behind a stagnation point in the domain. Some boundary conditions are applied to the new set of LCR equations. The work proceeds by discretization in space, which is based on the LBB-stable FE pair of $Q_2 P_1$ to maintain highly accurate solution. This is focused on a mixed FEM formulation of the velocity-pressure-stress that also deals with the LBB condition. In order to stabilize the chosen equal order interpolation between velocity and stress, two stabilization techniques are investigated. The first is by adding an artificial stress diffusion into the stress equation and the second deals with a jump stabilization over the element edge which is more elegant. The total resulting discrete system is treated by a monolithic technique. The design of this technique allows many other constitutive material laws to be easily integrated for later purposes. This is included inside an iterative Newton-multigrid approach. In one hand, the Newton is responsible for giving a quadratic convergence when the initial solution is close. On the other hand, the multigrid is responsible for solving the linearized algebraic system in an effective way. This approach is validated through a number of well-known benchmark problems. This includes the driven cavity, the flow around cylinder, MIT benchmark 2001, as well as the viscoelastic benchmark of flow around cylinder. Inside a driven cavity benchmark, the kinetic energy is calculated for high Re number. Here, the high order $Q_2 P_1$ pair is able to obtain results at Re = 10000. The flow around cylinder benchmark calculates drag and lift at Re = 20 and is meant to validate state of the art of the Newtonian flow within a few Newton steps. The MIT benchmark 2001 shows how the monolithic approach maintain the accuracy of time-dependent simulations. Here, hanging nodes are applied and show that they may increase accuracy of some functional values (in this case is the Nusselt number) if treated with care. The viscoelastic benchmark shows that the method is applicable as well to solve an additional coupled stress equation, and very important, it gives a high accurate results in comparison to many other approaches. After successful validations, few applications are presented in the area of generalized Newtonian flow as well as for viscoelastic flow.

The work shows that the presented numerical technique is applicable to any designed constitutive models given in any family of equation (1.4) as well as its LCR formulation if needed. When the flow models are hyperbolic PDE by nature (as in the viscoelastic case), there are some issues that appear in solving the aforementioned flow models within FEM. These are the reformulation of the original viscoelastic model that must be able to capture the high stress gradient, the choice of FEM functions that must follow the so-called LBB condition,

CHAPTER 7. SUMMARY AND OUTLOOK

the stabilization techniques if required and the robust outer/inner solvers. Unfortunately, most flow models are nonlinear in reality. And this is why this method is so meaningful. A strong solver alone may not be enough to solve the given flow models, and the stabilization technique must be able to cure unknown problems due to the mixed FEM or even due to the high nonlinearity appeared within the given model. The inner solver should be robust, which means clever enough, to find the direction of the global minima which is later controlled by a damping parameter from the outer solver. These components are then treated in a fully coupled way of solving in order to maintain accuracy of the solution at each time step or furthermore, in a direct steady simulation by just canceling the scaled mass matrix.

Finally, it is shown that a monolithic finite element framework with high order FEM (M-FEM) is very robust with respect to convergence and accuracy in dealing with the above mentioned constitutive models. Strong nonlinear and linear solvers are part of this robustness, which in this case are Newton (divided difference Jacobian) and multigrid (local MPSC smoother). A priori local refinement which leads to hanging nodes may increase accuracy in some cases if it is treated with care. The method seems to be a very promising way for highly nonlinear flow model or precisely in the case of solving viscoelastic flow for high We numbers with log-conformation representation (LCR). The advantages of using this method are its positivity preserving, no CFL-like restriction due to monolithic treatment, higher order with local adaptivity.

Future work could be on any directions, i.e. the implementation of LCR in other viscoelastic models that take the temperature effect into account, see [20], or in order to simulate more realistic flow problems, free surface technique with level set method or 3D extension of the current numerical method or a posteriori error control mechanisms for automatic grid refinement/coarsening. Other directions could be also parallelization of the method by introducing GPU computing, which means a new idea of dealing with local MPSC within multigrid approach or an in-depth study on the stabilization technique for high order element (EO-FEM) together with the choice of the interpolation functions.

Appendix

A.1 Derivation of LCR in details

Following Oldroyd, the general viscoelastic model is given by

$$\mathbf{T} + \Lambda \frac{\delta_a \mathbf{T}}{\delta t} = 2\eta_0 \left(\mathbf{D} + \Lambda_r \frac{\delta_a \mathbf{D}}{\delta t} \right) \quad \text{A.1}$$

where

$$\frac{\delta_a \mathbf{T}}{\delta t} = \frac{D\mathbf{T}}{Dt} + \frac{1-a}{2}(\nabla \mathbf{u} \cdot \mathbf{T} + \mathbf{T} \cdot \nabla \mathbf{u}^T) + \frac{1+a}{2}(-\mathbf{T} \cdot \nabla \mathbf{u} - \nabla \mathbf{u}^T \cdot \mathbf{T}) \quad \text{A.2}$$

$$\text{and} \quad \frac{D\mathbf{T}}{Dt} = \frac{\partial \mathbf{T}}{\partial t} + (\mathbf{u} \cdot \nabla)\mathbf{T} \quad \text{A.3}$$

By setting $a = 1$, the second term of the right hand side disappears,

$$\frac{\delta_1 \mathbf{T}}{\delta t} = \frac{\partial \mathbf{T}}{\partial t} + (\mathbf{u} \cdot \nabla)\mathbf{T} - \mathbf{T} \cdot \nabla \mathbf{u} - \nabla \mathbf{u}^T \cdot \mathbf{T}. \quad \text{A.4}$$

And in the same way,

$$\frac{\delta_1 \mathbf{D}}{\delta t} = \frac{\partial \mathbf{D}}{\partial t} + (\mathbf{u} \cdot \nabla)\mathbf{D} - \mathbf{D} \cdot \nabla \mathbf{u} - \nabla \mathbf{u}^T \cdot \mathbf{D} \quad \text{A.5}$$

Inserting these into equation A.1 gives

$$\mathbf{T} + \Lambda(\frac{\partial \mathbf{T}}{\partial t} + (\mathbf{u} \cdot \nabla)\mathbf{T} - \mathbf{T} \cdot \nabla \mathbf{u} - \nabla \mathbf{u}^T \cdot \mathbf{T}) = \\ 2\eta_0 \left(\mathbf{D} + \Lambda_r [\frac{\partial \mathbf{D}}{\partial t} + (\mathbf{u} \cdot \nabla)\mathbf{D} - \mathbf{D} \cdot \nabla \mathbf{u} - \nabla \mathbf{u}^T \cdot \mathbf{D}] \right) \quad \text{A.6}$$

Since the stress tensor consists of viscous and elastic part,

$$\mathbf{T} = 2\eta_0 \frac{\Lambda_r}{\Lambda} \mathbf{D} + \boldsymbol{\sigma}, \quad \text{A.7}$$

we can insert this into the left hand side of equation (A.6)

$$2\eta_0 \frac{\Lambda_r}{\Lambda}\mathbf{D} + \boldsymbol{\sigma} + \Lambda[\frac{\partial(2\eta_0 \frac{\Lambda_r}{\Lambda}\mathbf{D} + \boldsymbol{\sigma})}{\partial t} + (\mathbf{u} \cdot \nabla)(2\eta_0 \frac{\Lambda_r}{\Lambda}\mathbf{D} + \boldsymbol{\sigma}) \\ -(2\eta_0 \frac{\Lambda_r}{\Lambda}\mathbf{D} + \boldsymbol{\sigma}) \cdot \nabla \mathbf{u} - \nabla \mathbf{u}^T \cdot (2\eta_0 \frac{\Lambda_r}{\Lambda}\mathbf{D} + \boldsymbol{\sigma})] = \\ 2\eta_0 \left(\mathbf{D} + \Lambda_r [\frac{\partial \mathbf{D}}{\partial t} + (\mathbf{u} \cdot \nabla)\mathbf{D} - \mathbf{D} \cdot \nabla \mathbf{u} - \nabla \mathbf{u}^T \cdot \mathbf{D}] \right) \quad \text{A.8}$$

APPENDIX A. APPENDIX

We rearrange the last equation,

$$2\eta_0 \frac{\Lambda_r}{\Lambda} \mathbf{D} + \boldsymbol{\sigma} +$$
$$+ 2\eta_0 \frac{\Lambda_r}{\Lambda} \Lambda \left(\frac{\partial \mathbf{D}}{\partial t} + (\mathbf{u} \cdot \nabla)\mathbf{D} - \mathbf{D} \cdot \nabla \mathbf{u} - \nabla \mathbf{u}^T \cdot \mathbf{D} \right)$$
$$+ \Lambda \left(\frac{\partial \boldsymbol{\sigma}}{\partial t} + (\mathbf{u} \cdot \nabla)\boldsymbol{\sigma} - \nabla \mathbf{u} \cdot \boldsymbol{\sigma} - \boldsymbol{\sigma} \cdot \nabla \mathbf{u}^T \right) = \quad \text{A.9}$$
$$2\eta_0 \mathbf{D} + 2\eta_0 \Lambda_r \left(\frac{\partial \mathbf{D}}{\partial t} + (\mathbf{u} \cdot \nabla)\mathbf{D} - \mathbf{D} \cdot \nabla \mathbf{u} - \nabla \mathbf{u}^T \cdot \mathbf{D} \right)$$

Now, we can see that the second term of the left and right hand side cancel out,

$$2\eta_0 \frac{\Lambda_r}{\Lambda} \mathbf{D} + \boldsymbol{\sigma} +$$
$$+ \Lambda \left(\frac{\partial \boldsymbol{\sigma}}{\partial t} + (\mathbf{u} \cdot \nabla)\boldsymbol{\sigma} - \nabla \mathbf{u} \cdot \boldsymbol{\sigma} - \boldsymbol{\sigma} \cdot \nabla \mathbf{u}^T \right) = \quad \text{A.10}$$
$$2\eta_0 \mathbf{D}$$

By introducing $\beta = \frac{\Lambda_r}{\Lambda}$, we end up with Oldroyd-B model which is given in elastic stress equation,

$$\Lambda \left(\frac{\partial \boldsymbol{\sigma}}{\partial t} + (\mathbf{u} \cdot \nabla)\boldsymbol{\sigma} - \nabla \mathbf{u} \cdot \boldsymbol{\sigma} - \boldsymbol{\sigma} \cdot \nabla \mathbf{u}^T \right) + 2\eta_0 \left(\beta - 1 \right) \mathbf{D} + \boldsymbol{\sigma} = 0. \quad \text{A.11}$$

A.1.1 Oldroyd-B in conformation tensor formulation

Before we describe LCR, we need to reformulate equation (A.11) into its conformation stress tensor formulation. First, we introduce the conformation stress tensor,

$$\boldsymbol{\sigma} = \frac{\eta_p}{\Lambda}(\boldsymbol{\tau} - \mathbf{I}). \quad \text{A.12}$$

By inserting this into equation (A.11), we have

$$\Lambda \left(\frac{\partial (\frac{\eta_p}{\Lambda}(\boldsymbol{\tau} - \mathbf{I}))}{\partial t} + (\mathbf{u} \cdot \nabla)(\frac{\eta_p}{\Lambda}(\boldsymbol{\tau} - \mathbf{I}) - \nabla \mathbf{u} \cdot (\frac{\eta_p}{\Lambda}(\boldsymbol{\tau} - \mathbf{I}) - (\frac{\eta_p}{\Lambda}(\boldsymbol{\tau} - \mathbf{I}) \cdot \nabla \mathbf{u}^T \right)$$
$$+ 2\eta_0 \left(\beta - 1 \right) \mathbf{D} + \frac{\eta_p}{\Lambda}(\boldsymbol{\tau} - \mathbf{I}) = 0. \quad \text{A.13}$$

Since the time derivative of unity and the convective part of it are zero, we can rewrite the last equation to be,

$$\Lambda \left(\frac{\eta_p}{\Lambda} \frac{\partial \boldsymbol{\tau}}{\partial t} + \frac{\eta_p}{\Lambda}(\mathbf{u} \cdot \nabla)\boldsymbol{\tau} - \frac{\eta_p}{\Lambda}(\nabla \mathbf{u} \cdot \boldsymbol{\tau} - \nabla \mathbf{u} \cdot \mathbf{I}) - \frac{\eta_p}{\Lambda}(\boldsymbol{\tau} \cdot \nabla \mathbf{u}^T - \mathbf{I} \cdot \nabla \mathbf{u}^T) \right)$$
$$+ 2\eta_0 \left(\beta - 1 \right) \mathbf{D} + \frac{\eta_p}{\Lambda}(\boldsymbol{\tau} - \mathbf{I}) = 0 \quad \text{A.14}$$

Here, one remembers that $\nabla \mathbf{u} \cdot \mathbf{I} + \mathbf{I} \cdot \nabla \mathbf{u}^T = 2\mathbf{D}$. Hence,

$$\eta_p \left(\frac{\partial \boldsymbol{\tau}}{\partial t} + (\mathbf{u} \cdot \nabla)\boldsymbol{\tau} - \nabla \mathbf{u} \cdot \boldsymbol{\tau} - \boldsymbol{\tau} \cdot \nabla \mathbf{u}^T \right)$$
$$+ 2\eta_p \mathbf{D} + 2\eta_0 \left(\beta - 1 \right) \mathbf{D} + \frac{\eta_p}{\Lambda}(\boldsymbol{\tau} - \mathbf{I}) = 0 \quad \text{A.15}$$

A.2. DERIVATION OF CONFORMATION TENSOR INFLOW CONDITIONS

Since $\eta_p = \eta_0 - \eta_0 \beta$, the middle terms of the left hand side cancel out. Thus, the Oldroyd-B can be rewritten in terms of conformation stress tensor reformulation,

$$\frac{\partial \boldsymbol{\tau}}{\partial t} + \overbrace{(\mathbf{u} \cdot \nabla) \boldsymbol{\tau}}^{\text{convection}} \underbrace{- \nabla \mathbf{u} \cdot \boldsymbol{\tau} - \boldsymbol{\tau} \cdot \nabla \mathbf{u}^T}_{\text{stretching}} + \frac{1}{\Lambda}(\boldsymbol{\tau} - \mathbf{I}) = 0. \qquad \text{A.16}$$

A.1.2 Oldroyd-B in LCR formulation

Lcr introduces a new variable $\boldsymbol{\psi} = R \, log(\boldsymbol{\tau}) R^T$ through eigenvalue computations, which is placed on every cubature points inside the FE code. The reformulation starts with rotating the conformation tensor $\boldsymbol{\tau}$ into its main principle axis (diagonalization process)

$$R^T \boldsymbol{\tau} R = \text{diag}(\lambda_1, \lambda_2) \qquad \text{A.17}$$

with R being an orthogonal matrix, and apply the same rotation to the velocity gradient,

$$\nabla \mathbf{u} = \mathbf{B} + \boldsymbol{\Omega} + \mathbf{N} \boldsymbol{\tau}^{-1} \qquad \text{A.18}$$

$$R^T \nabla \mathbf{u} R = R^T \mathbf{B} R + R^T \boldsymbol{\Omega} R + R^T \mathbf{N} \boldsymbol{\tau}^{-1} R. \qquad \text{A.19}$$

The goal is to design a symmetric matrix $\mathbf{B}$, which commutes with the conformation tensor. Having this in mind, we can easily set $R^T \mathbf{B} R = \text{diag}(R^T \nabla \mathbf{u} R)$. Thus, $R^T \nabla \mathbf{u} R = \begin{pmatrix} m_{11} & m_{12} \\ m_{21} & m_{22} \end{pmatrix}$, $R^T \mathbf{B} R = \begin{pmatrix} m_{11} & 0 \\ 0 & m_{22} \end{pmatrix}$ while the matrices $\boldsymbol{\Omega}, \mathbf{N}$ are set to be pure rotational, $R^T \boldsymbol{\Omega} R = \begin{pmatrix} 0 & \omega \\ -\omega & 0 \end{pmatrix}$, $R^T \mathbf{N} R = \begin{pmatrix} 0 & n \\ -n & 0 \end{pmatrix}$. By inserting all matrices into equation (A.19), we can easily determine the component of matrices $\boldsymbol{\Omega}, \mathbf{N}$ to be, $n = \frac{m_{12} + m_{21}}{\lambda_2^{-1} - \lambda_1^{-1}}$ and $\omega = \frac{\lambda_2 m_{12}}{\lambda_1 m_{21}}$.

Now, by inserting equation (A.18) into the stretching part of equation (A.16) gives

$$- \nabla \mathbf{u} \cdot \boldsymbol{\tau} - \boldsymbol{\tau} \cdot \nabla \mathbf{u}^T = \qquad \text{A.20}$$

$$-(\mathbf{B} + \boldsymbol{\Omega} + \mathbf{N} \boldsymbol{\tau}^{-1}) \cdot \boldsymbol{\tau} - \boldsymbol{\tau} \cdot (\mathbf{B} + \boldsymbol{\Omega} + \mathbf{N} \boldsymbol{\tau}^{-1})^T \qquad \text{A.21}$$

Now, remember that $\mathbf{B} \cdot \boldsymbol{\tau} = \boldsymbol{\tau} \cdot \mathbf{B}$ (commutable) and $\boldsymbol{\Omega}, \mathbf{N}$ are pure rotational matrices $\mathbf{N} = -\mathbf{N}^T$, we obtain the stretching part to be

$$- \nabla \mathbf{u} \cdot \boldsymbol{\tau} - \boldsymbol{\tau} \cdot \nabla \mathbf{u}^T = -2\mathbf{B} \cdot \boldsymbol{\tau} - (\boldsymbol{\Omega} \boldsymbol{\tau} - \boldsymbol{\tau} \boldsymbol{\Omega}) \qquad \text{A.22}$$

and equation (A.16) evolves into

$$\frac{\partial \boldsymbol{\tau}}{\partial t} + (\mathbf{u}.\nabla)\boldsymbol{\tau} - (\boldsymbol{\Omega}.\boldsymbol{\tau} - \boldsymbol{\tau}.\boldsymbol{\Omega}) - 2\mathbf{B}.\boldsymbol{\tau} = \frac{1}{\Lambda}(\mathbf{I} - \boldsymbol{\tau}). \qquad \text{A.23}$$

Finally, by replacing the conformation tensor with the new variable $\boldsymbol{\psi} = log(\boldsymbol{\tau})$, the Oldroyd-B model evolves again into

$$\frac{\partial \boldsymbol{\psi}}{\partial t} + (\mathbf{u}.\nabla)\boldsymbol{\psi} - (\boldsymbol{\Omega}.\boldsymbol{\psi} - \boldsymbol{\psi}.\boldsymbol{\Omega}) - 2\mathbf{B} = \frac{1}{\Lambda}\left(e^{-\boldsymbol{\psi}} - \mathbf{I}\right). \qquad \text{A.24}$$

A.2 Derivation of conformation tensor inflow conditions

The viscoelastic fluid has a memory during motion, which means its stress components are not zero at the inflow, when we consider a periodic flow motion. The stress inflow condition

APPENDIX A. APPENDIX

is derived from the fully developed velocity inflow $u_x(y)$ by solving the constitutive material law in equation (A.16). The fully developed inflow velocity means that at the inflow $u_y = 0$, $\frac{\partial u_x}{\partial x} = 0$. From continuity condition, it also implies that $\frac{\partial u_y}{\partial y} = 0$. We start with recalling equation (A.16),

$$\frac{\partial \boldsymbol{\tau}}{\partial t} + (\mathbf{u} \cdot \nabla)\boldsymbol{\tau} - \nabla \mathbf{u} \cdot \boldsymbol{\tau} - \boldsymbol{\tau} \cdot \nabla \mathbf{u}^T + \frac{1}{\Lambda}(\boldsymbol{\tau} - \mathbf{I}) = 0 \qquad \text{A.25}$$

and by dropping time derivative and the convection, which vanishes on the inflow, we obtain

$$-\nabla \mathbf{u} \cdot \boldsymbol{\tau} - \boldsymbol{\tau} \cdot \nabla \mathbf{u}^T + \frac{1}{\Lambda}(\boldsymbol{\tau} - \mathbf{I}) = 0 \qquad \text{A.26}$$

$$\text{A.27}$$

We insert fully developed inflow assumptions,

$$-\begin{pmatrix} 0 & \frac{\partial u_x}{\partial y} \\ 0 & 0 \end{pmatrix}\begin{pmatrix} \tau_{xx} & \tau_{xy} \\ \tau_{yx} & \tau_{yy} \end{pmatrix} - \begin{pmatrix} \tau_{xx} & \tau_{xy} \\ \tau_{yx} & \tau_{yy} \end{pmatrix}\begin{pmatrix} 0 & 0 \\ \frac{\partial u_x}{\partial y} & 0 \end{pmatrix} \qquad \text{A.28}$$

$$+\frac{1}{\Lambda}\left[\begin{pmatrix} \tau_{xx} & \tau_{xy} \\ \tau_{yx} & \tau_{yy} \end{pmatrix} - \begin{pmatrix} 1 & 0 \\ 0 & 1 \end{pmatrix}\right] = 0 \qquad \text{A.29}$$

$$\text{A.30}$$

Since conformation stress tensor is symmetric matrix, we have 3 linear scalar systems,

$$\frac{1}{\Lambda}(1 - \tau_{yy}) = 0 \qquad \text{A.31}$$

$$-\tau_{yy}\frac{\partial u_x}{\partial y} = \frac{1}{\Lambda}(0 - \tau_{xy}) \qquad \text{A.32}$$

$$-2\frac{\partial u_x}{\partial y}\tau_{xy} = \frac{1}{\Lambda}(1 - \tau_{xx}) \qquad \text{A.33}$$

These equations yield

$$\tau_{yy} = 1, \quad \tau_{xy} = \Lambda\frac{\partial u_x}{\partial y}, \quad \tau_{xx} = 1 + 2\left(\Lambda\frac{\partial u_x}{\partial y}\right)^2 \qquad \text{A.34}$$

at the inflow.

A.3 Eigenvalues implementations

LCR needs eigenvalues computation for the transformation from the numerical variable, $\boldsymbol{\psi}$, into the conformation stress tensor, $\boldsymbol{\tau}$. The exponential relation between the two variables are laid on the eigenvalues,

$$\boldsymbol{\psi} = \begin{pmatrix} c & s \\ -s & c \end{pmatrix}\begin{pmatrix} \lambda_1 & 0 \\ 0 & \lambda_2 \end{pmatrix}\begin{pmatrix} c & -s \\ s & c \end{pmatrix} \qquad \text{A.35}$$

$$\boldsymbol{\psi} = \begin{pmatrix} c^2\lambda_1 - s^2\lambda_2 & cs\lambda_1 + cs\lambda_2 \\ -cs\lambda_1 - cs\lambda_2 & c^2\lambda_2 - s^2\lambda_1 \end{pmatrix} \qquad \text{A.36}$$

where $c^2 + s^2 = 1$. This looks very easy. Yet, inside the numerical code, we must be careful of division by zero that might appear from approximation of $\boldsymbol{\psi}$ in each time step. To avoid this, the rotation matrix components can be derived from equation (A.36) by setting some if-condition.

A.4. DETAILS OF STRETCHING IN THE WAKE REGION

If $\lambda_2 = -\lambda_1$ then

$$c^2 = \frac{1}{2} + \frac{1}{4}\frac{\psi_{11}-\psi_{22}}{\lambda_1}, \quad s^2 = \frac{1}{2} + \frac{1}{4}\frac{\psi_{22}-\psi_{11}}{\lambda_1}, \quad cs = -\frac{1}{2}\frac{\psi_{12}}{\lambda_1}. \qquad \text{A.37}$$

If $\lambda_2 = \lambda_1$ then this is a rare situation which can only happen in the case of $\lambda_2 = \lambda_1 = 0$ in the first iteration when $\boldsymbol{\psi} = 0$ and the conformation stress tensor is equal to unity, $\boldsymbol{\tau} = \mathbf{I}$.
If $\lambda_2 \ne \lambda_1$ then this is the most cases when numerical iterations start,

$$c^2 = \frac{\psi_{11}\lambda_1 - \psi_{22}\lambda_2}{\lambda_1^2 - \lambda_2^2}, \quad s^2 = \frac{\psi_{11}\lambda_2 - \psi_{22}\lambda_1}{\lambda_2^2 - \lambda_1^2}, \quad cs = \frac{\psi_{12}}{\lambda_2 - \lambda_1} \qquad \text{A.38}$$

A.4 Details of stretching in the wake region

In Chapter 5 we learn that Oldroyd-B fluid suffers from uniaxial extensional flow in the symmetry line behind the cylinder (stagnation point). Here, we will revisit the Oldroyd-B with conformation stress tensor formulation to take a closer look what could possibly happen in the wake region,

$$\frac{\partial \boldsymbol{\tau}}{\partial t} + \overbrace{(\mathbf{u}\cdot\nabla)\boldsymbol{\tau}}^{\text{convection}} \underbrace{-\nabla\mathbf{u}\cdot\boldsymbol{\tau} - \boldsymbol{\tau}\cdot\nabla\mathbf{u}^T}_{\text{stretching}} + \frac{1}{\Lambda}(\boldsymbol{\tau}-\mathbf{I}) = 0. \qquad \text{A.39}$$

In the symmetry line, we are interested in the axial conformation stress tensor, τ_{11},

$$\frac{\partial \tau_{11}}{\partial t} + u_x\frac{\partial \tau_{11}}{\partial x} + u_y\frac{\partial \tau_{11}}{\partial x} - \frac{\partial u_x}{\partial x}\tau_{11} - \tau_{11}\frac{\partial u_x}{\partial x} + \frac{1}{\Lambda}(\tau_{11}-1) = 0. \qquad \text{A.40}$$

Now, we can cancel the time derivative for steady solution and $u_y = 0$ along the symmetry line,

$$u_x\frac{\partial \tau_{11}}{\partial x} - 2\frac{\partial u_x}{\partial x}\tau_{11} + \frac{1}{\Lambda}(\tau_{11}-1) = 0 \quad \text{or}$$

$$\frac{\partial \tau_{11}}{\partial x} = \frac{2}{u_x}\frac{\partial u_x}{\partial x}\tau_{11} - \frac{1}{u_x\Lambda}(\tau_{11}-1) \qquad \text{A.41}$$

Since we are interested in the maximum value of τ_{11}, the first derivative should vanish,

$$0 = \frac{2}{u_x}\frac{\partial u_x}{\partial x}\tau_{11} - \frac{1}{u_x\Lambda}(\tau_{11}-1) \quad \text{or}$$

$$\text{Max. } \tau_{11} = \frac{1}{1 - 2\Lambda\frac{\partial u_x}{\partial x}} \qquad \text{A.42}$$

This suggests that in the case of Oldroyd-B model, $\Lambda\frac{\partial u_x}{\partial x} \ne \frac{1}{2}$, otherwise conformation stress tensor value is not physically reasonable. In the case of Giesekus, there is an additional second order conformation stress tensor that bounds the maximum stretching behavior to be somehow finite. Having in mind that $\tau_{12} = 0$ in the symmetry line and mobility factor is $\alpha < 1$, the same derivations as above for Giesekus gives

$$-\alpha\tau_{11}^2 - \left(1 - 2\alpha - 2\Lambda\frac{\partial u_x}{\partial x}\right)\tau_{11} + (1-\alpha) = 0 \quad \text{or}$$

$$\tau_{11} = \frac{\left(1 - 2\alpha - 2\Lambda\frac{\partial u_x}{\partial x}\right) \pm \sqrt{\left(1 - 2\alpha - 2\Lambda\frac{\partial u_x}{\partial x}\right)^2 + 4\alpha(1-\alpha)}}{-2\alpha} \qquad \text{A.43}$$

which suggests no division by zero nor any imaginary number when $\alpha \ne 0$.

APPENDIX A. APPENDIX

A.5 More MIT Benchmark 2001 results

Additional MIT Benchmark 2001 results in plots of periodical oscillations of velocity at point 1 and the Nusselt number.

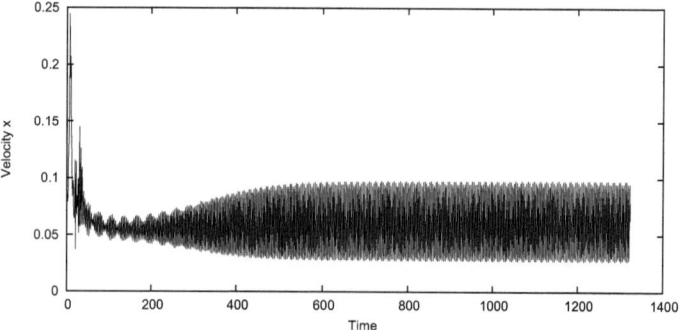

Figure A.1: Velocity x oscillations at point 1

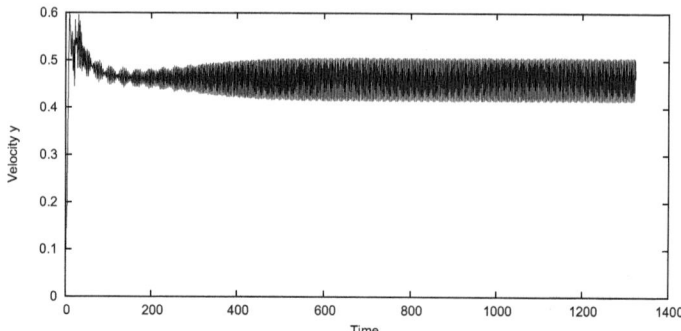

Figure A.2: Velocity y oscillations at point 1

A.5. MORE MIT BENCHMARK 2001 RESULTS

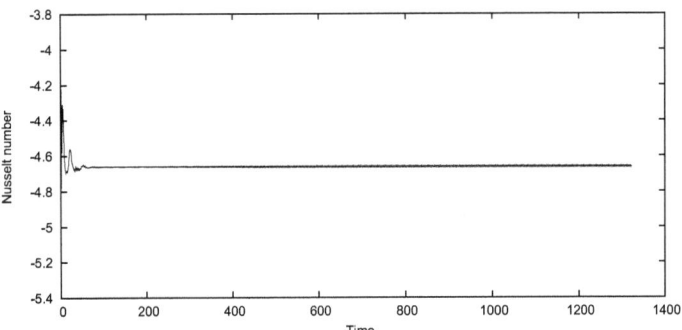

Figure A.3: Nusselt number oscillations

Bibliography

[1] Afonso, A., Oliveira, P. J., Pinho, F. T., and Alves, M. A. The log-conformation tensor approach in the finite-volume method framework. *Journal of Non-Newtonian Fluid Mechanics*, 157:55–65, 2009.

[2] Arnold, D.N., Boffi, D., and Falk, R.S. Approximation by Quadrilateral Finite Elements. *Math. Comput.*, 71(239):909–922, 2002.

[3] Baaijens, F.P.T. Mixed finite element methods for viscoelastic flow analysis: a review. *J. Non-Newtonian Fluid Mechanics*, 79:361–385, 1998.

[4] Baaijens, F.P.T., Selen, S.H.A, Baaijens, H.P.W., and Meijer, H.E.H. Viscoelastic flow past a confined cylinder of a low density polyethylene melt. *J. Non-Newtonian Fluid Mechanics*, 68:173–203, 1997.

[5] Baranger, J. and Sandri, D. Formulation of Stokes problem and the linear elasticity equations suggested by Oldroyd model for viscoelastic flow. *Math. Modell. Numer. Anal*, 26:231–345, 1992.

[6] Bathe, K., J., Editor. *Computational Fluid and Solid Mechanics*. Elsevier, 2001. Proceedings First MIT Conference on Computational Fluid and Solid Mechanics Vol. 2.

[7] Behr M. *Stabilized finite element methods for incompressible flows with emphasis on moving boundaries and interfaces*. PhD thesis, University of Minnesota, Department of Aerospace Engineering and Mechanics, 1992.

[8] Beris, A. N., Armstrong, R. C., and Brown, R. A. Finite element calculation of viscoelastic flow in a journal bearing: I, small eccentries. *Journal of non-Newtonian Fluid Mechanics*, 16:141–172, 1984.

[9] Beris, A. N., Armstrong, R. C., and Brown, R. A. Finite element calculation of viscoelastic flow in a journal bearing: Ii, moderate eccentries. *Journal of non-Newtonian Fluid Mechanics*, 19:323–347, 1986.

[10] Boffi, D. and Gastaldi, L. On the quadrilateral Q2-P1 element for the Stokes problem. *Int. J. Numer. Meth. Fluids*, 39:1001–1011, 2002.

[11] Boger, D. V. and Walters, K. *Rheological Phenomena in Focus*. Elsevier, 1993. Amsterdam.

[12] Bonito, A. and Burman, E. A continuous interior penalty method for viscoelastic flows. *Siam Journal of Scientific Computing*, 30:1156–1177, 2008.

[13] Brezzi, F. and Fortin, M. *Mixed and Hybrid Finite Element methods*. Springer, Berlin, 1986.

[14] Brooks, A. N. and Hughes, T. J. R. Streamline upwind / petrov-galerkin formulations for convection dominated flows with particular emphasis on the incompressible navier-stokes equations. *Comput. Methods Appl. Mech. Eng.*, 32:199–259, 1982.

[15] Bänsch, E. Local mesh refinement in 2 and 3 dimension. *IMPACT of Computing in Science and Engineering*, 3:181–191, 1991.

[16] Christon, M. A., Gresho, P. M., and Sutton, S. B. Computational predictability of natural convection flows in enclosures. *In: Computational Fluid and Solid Mechanics*, 40:1465–1468, 2001. First MIT Conference on Computational Fluid and Solid Mechanics.

BIBLIOGRAPHY

[17] Christon, M. A., Gresho, P. M., and Sutton, S. B. Computational predictability of natural convection flows in enclosures. *International Journal for Numerical Methods in Fluids*, 40:953–980, 2002.

[18] Coronado, O. M., Arora, D., Behr, M., and Pasqualli, M. A simple method for simulating general viscoelastic fluid flows with an alternate log-conformation formulation. *Journal of Non-Newtonian Fluid Mechanics*, 147:189–199, 2007.

[19] Crochet, M. J. and Bezy, M. Numerical solution for the flow of viscoelastic fluids. *Journal of non-Newtonian Fluid Mechanics*, 5:201–218, 1979.

[20] Damanik, H., Hron, J., Ouazzi, A., and Turek, S. A monolithic FEM–multigrid solver for non-isothermal incompressible flow on general meshes. *Journal of Computational Physics*, 228:3869–3881, 2009.

[21] Damanik, H., Hron, J., Ouazzi, A., and Turek, S. A monolithic FEM approach for the log-conformation reformulation (lcr) of viscoelastic flow problems. *Journal of non-Newtonian Fluid Mechanics*, 165:1105–1113, 2010.

[22] Davis, D. and Bänsch, E. An operator-splitting finite-element approach to the 8:1 thermal cavity problem. *International Journal for Numerical Methods in Fluids*, 40:1019–1030, 2002.

[23] Davis, T. A. Algorithm 832: Umfpack, an unsymmetric-pattern multifrontal method. *ACM Trans. Math. Softw.*, 30(2):196–199, 2004.

[24] Dennis, Jr., J. E. and Schnabel, R. B. *Numerical Methods for Unconstrained Optimization and Nonlinear Equations*. SIAM, 1996.

[25] Dou, H.-S. and Phan-Thien, N. The flow of an Oldroyd–B fluid past a cylinder in a channel: adaptive viscosity vorticity (davss-ω) formulation. *Journal of Non-Newtonian Fluid Mechanics*, 87:47–73, 1999.

[26] Fan, Y., Tanner, R. I.., and Phan-Thien, N. Galerkin/least-square finite element methods for steady viscoelastic flows. *Journal of Non-Newtonian Fluid Mechanics*, 84:233–256, 1999.

[27] Fan, Y., Yang, H., and Tanner, R. I. Stress boundary layers in the viscoelastic flow past a cylinder in a channel: limiting solutions. *Acta Mech Sinica*, 21:311–321, 2005.

[28] Fattal, R. and Kupferman, R. Constitutive laws for the matrix-logarithm of the conformation tensor. *J. Non-Newton. Fluid Mech.*, 123:281–285, 2004.

[29] Fortin, A. and Guénette, R.and Pierre R. On the Discrete EVSS Method. *Comp. Methods Appl. Mech. Engrg.*, 189:121–139, 2000.

[30] Fortin, M. and Fortin, A. A new approach for the FEM simulation of viscoelastic flows. *J. Non-Newtonian Fluid Mech*, pages 295–310, 1989.

[31] Fortin, M., Guenette, R., and Pierre, R. Numerical analysis of the modified EVSS method. *Computer methods in applied mechanics and engineering*, 143:79–95, 1997.

[32] Fortin, M. and Pierre, R. On the convergence of the mixed method of Crochet and Marchal for viscoelastic flows, Computer Methods in Applied. *Mechanics and Engineering*, 73:341–350, 1989.

BIBLIOGRAPHY

[33] Giesekus, H. A simple constitutive equation for polymer fluids based on the concept of deformation-dependent tensorial mobility. *Journal of non-Newtonian Fluid Mechanics*, 11:69–109, 1982.

[34] Girault, V. and Raviart, P. A. *Finite Element Methods for Navier-Stokes equations*. Springer, 1986. Berlin-Heidelberg.

[35] Gresho, P. M. and Chan, S. T. On the theory of semi-implicit projection methods for viscous incompressible flow and its implementation via a finite element method that also introduces a nearly consistent mass matrix. ii: Implementation. *International Journal for Numerical Methods in Fluids*, 11(5):621–659, 1990.

[36] Gresho, P. M. and Sutton, S. B. Application of the fidap code to the 8:1 thermal cavity problem. *International Journal for Numerical Methods in Fluids*, 40:1083–1092, 2002.

[37] Heywood, J. G., Rannacher, R., and Turek, S. Artificial boundaries and flux and pressure conditions for the incompressible Navier-Stokes equations. *International Journal for Numerical Methods in Fluids*, 22:325–352, 1992.

[38] Hron, J. and Turek, S. A monolithic FEM/multigrid solver for ALE formulation of fluid structure interaction with application in biomechanics. In H.-J. Bungartz and M. Schäfer, editors, *Fluid-Structure Interaction – Modelling, Simulation, Optimization*, number 53 in Lecture Notes in Computational Science and Engineering, pages 146–170. Springer, Berlin, 2006. ISBN 3-540-34595-7.

[39] Hughes, T. J. R., Franca, L. P., and Hulbert G. M. A new finite element formulation for computational fluid dynamics, viii: the galerkin/least-squares method for advective-diffusive equations. *Comput. Methods Appl. Mech. Eng.*, 73:173–189, 1989.

[40] Hulsen, M. A. A sufficient condition for a positive definite configuration tensor in differential models. *Journal of Non-Newtonian Fluid Mechanics*, 38:93–100, 1990.

[41] Hulsen, M. A., Fattal, R., and Kupferman, R. Flow of viscoelastic fluids past a cylinder at high Weissenberg number: Stabilized simulations using matrix logarithms. *Journal of Non-Newtonian Fluid Mechanics*, 127:27–39, 2005.

[42] Ii, S., Sugiyama, K., Takeuchi, S., Takagi, S., and Matsumoto, Y. An implicit full eulerian method for the fluid-structure interaction problem. *International Journal for Numerical Methods in Fluids*, 65:150–165, 2010.

[43] John, V. Higher order finite element methods and multigrid solvers in a benchmark problem for the 3-D Navier-Stokes equations. *IJNMF*, 40:775–798, 2002.

[44] Joseph, D. D. *Fluid Dynamics of Viscoelastic Liquids*. Springer, 1990. Applied Mathematical Sciences 84.

[45] Keunings, R. A survey of computational rheology. *Proc. XIIIth Int. Congr. on Rheology*, 1:7–14, 2000.

[46] Larson, R. G. . *The Structure and Rheology of Complex Fluids*. Oxford University Press, 1999. .

[47] Layton, W. J. *Introduction to the numerical analysis of incompressible viscous flows*. SIAM, 2008.

[48] Lee, Y.-J. and Xu, J. New formulations, positivity preserving discretizations and stability analysis for non-newtonian flow models. *Computer methods in applied mechanics and engineering*, 195:1180–1206, 2006.

BIBLIOGRAPHY

[49] Luo, X.-L. and Tanner, R. I. A streamline element scheme for solving viscoelastic flow problems. part i. differential constitutive models. *Journal of Non-Newtonian Fluid Mechanics*, 21:179–199, 1986.

[50] Luo, X.-L. and Tanner, R. I. A streamline element scheme for solving viscoelastic flow problems. part ii. integral constitutive models. *Journal of Non-Newtonian Fluid Mechanics*, 22:61–89, 1986.

[51] Marchal, J.M. and Crochet, M.J. A new mixed finite element for calculating viscoelastic flow. *J. Non-Newtonian. Fluid Mech*, 26:77–114, 1987.

[52] Meyer A. Projected pcgm for handling hanging nodes in adaptive finite element procedures. Preprint sfb sfb393/99-25, TU Chemnitz, 1999.

[53] Niemunis, A. *Extended Hypoplastic models for soils*. Schriftreihe des Inst. f. Grundbau u. Bodenmechanik der Ruhr-Universitaet Bochum, 2003. Heft 34.

[54] Oldroyd, J. G. On the formulation of rheological equations of state. *Proc. R. Soc. London, Ser. A* 200:523–541, 1950.

[55] A. Ouazzi and S. Turek. Efficient multigrid and data structures for edge–oriented FEM stabilization. In *Numerical Mathematics and Advanced Applications Enumath 2005*, pages 520–527. Springer, Berlin, 2006. ISBN-10 3-540-34287-7.

[56] Ouazzi, A. *Finite Element Simulation of Nonlinear Fluids. Application to Granular Material and Powder*. Shaker Verlag, Aachen, Germany, 2006. ISBN 3-8322-5201-0.

[57] Owens, R. G. and Phillips, T. N. *Computational Rheology*. Imperial College Press, 2002. London.

[58] Pan, T.-W. and Hao, J. Numerical simulation of a lid-driven cavity viscoelastic flow at high Weissenberg numbers. *C. R. Acad. Sci. Paris*, 344:283–286, 2007.

[59] Press, W. H., Teukolsky, S., Vetterling, W. T., and Flannery, B. P. *Numerical Recipes in C++. The Art of Scientific Computing*. Cambridge University Press, 2002. ISBN 0-521-75033-4.

[60] Rannacher, R. and Turek, S. A Simple nonconforming quadrilateral Stokes element. *Numer. Methods Partial Differential Equations*, 8:97–111, 1992.

[61] Renardy, M. *Mathematical Analysis of Viscoelastic Flows*. SIAM, 2000. CBMS-NSF Regional Conference Series in Applied Mathematics.

[62] Schmachtel, R. and Turek, S. Fully coupled and operator–splitting approaches for natural convection. *International Journal for Numerical Methods in Fluids*, 40:1109–1119, 2002.

[63] Schowalter, W. R. *Mechanics of Non-Newtonian Fluids*. Pergamon Press, 1978. Great Britain.

[64] Shihe Xin and Le Quéré, P. An extended chebyshev pseudo-spectral benchmark for the 8:1 differentially heated cavity. *International Journal for Numerical Methods in Fluids*, 40:981–998, 2002.

[65] Sobotka, Z. *Rheology of Materials and Engineering Structures*. Academia, Prague, 1984.

[66] Sureshkumar, R. and Beris, N. Effect of artificial stress diffusivity on the stability of numerical calculations and the flow dynamics of time-dependent viscoelastic flows. *Journal of non-Newtonian Fluid Mechanics*, 60:53–80, 1995.

BIBLIOGRAPHY

[67] S. Turek, A. Ouazzi, and J. Hron. On pressure separation algorithms (PSEPA) for improving the accuracy of incompressible flow simulations. *International Journal for Numerical Methods in Fluids*, 59(4):387–403, 2008.

[68] Turek, S. A comparative study of time–stepping techniques for the incompressible Navier–Stokes equations: From fully implicit non–linear schemes to semi–implicit projection methods. *International Journal for Numerical Methods in Fluids*, 22:987–1011, 1996.

[69] Turek, S. *Efficient solvers for incompressible flow problems: An algorithmic and computational approach.* Springer, 1999. LNCSE 6.

[70] Turek, S. and Hron, J. A monolithic FEM solver for an ALE formulation of fluid–structure interaction with configuration for numerical benchmarking. In P. Wesseling, E. Onate, and J. Periaux, editors, *Books of Abstracts European Conference on Computational Fluid Dynamics*, page 176. nn, 2006. Eccomas CFD 2006.

[71] Turek, S. and Ouazzi, A. Unified edge–oriented stabilization of nonconforming FEM for incompressible flow problems: Numerical investigations. *J. Numer. Math.*, 15:299–322, 2007.

[72] Turek, S., Ouazzi, A., and Schmachtel, R. Multigrid Methods for Stabilized Nonconforming Finite Elements for Incompressible Flow involving the Deformation Tensor formulation. *J. Numer. Math.*, 10:235–248, 2002.

[73] Turek, S. and Schäfer, M. Benchmark computations of laminar flow around cylinder. In E. H. Hirschel, editor, *Flow Simulation with High-Performance Computers II*, volume 52 of *Notes on Numerical Fluid Mechanics*, pages 547–566. Vieweg, 1996. co. F. Durst, E. Krause, R. Rannacher.

[74] Verfürth, R. *A review of a posteriori error estimation and adaptive mesh refinement techniques.* Wiley and Teubner, 1996.

[75] Wesseling, P. *An Introduction to Multigrid Methods.* John Wiley & Sons, 1992.

[76] Yoo, J. Y. and Na, Y. A numerical study of the planar contraction flow of a viscoelastic fluid using the simpler algorithm. *Journal of Non-Newtonian Fluid Mechanics*, 30:89–106, 1991.

Die VDM Verlagsservicegesellschaft sucht für wissenschaftliche Verlage abgeschlossene und herausragende

Dissertationen, Habilitationen, Diplomarbeiten, Master Theses, Magisterarbeiten usw.

für die kostenlose Publikation als Fachbuch.

Sie verfügen über eine Arbeit, die hohen inhaltlichen und formalen Ansprüchen genügt, und haben Interesse an einer honorarvergüteten Publikation?

Dann senden Sie bitte erste Informationen über sich und Ihre Arbeit per Email an *info@vdm-vsg.de*.

Sie erhalten kurzfristig unser Feedback!

VDM Verlagsservicegesellschaft mbH
Dudweiler Landstr. 99
D - 66123 Saarbrücken

Telefon +49 681 3720 174
Fax   +49 681 3720 1749

www.vdm-vsg.de

Die VDM Verlagsservicegesellschaft mbH vertritt

Printed by Books on Demand GmbH, Norderstedt / Germany